Springer Theses

Recognizing Outstanding Ph.D. Research

For further volumes:
http://www.springer.com/series/8790

Aims and Scope

The series "Springer Theses" brings together a selection of the very best Ph.D. theses from around the world and across the physical sciences. Nominated and endorsed by two recognized specialists, each published volume has been selected for its scientific excellence and the high impact of its contents for the pertinent field of research. For greater accessibility to non-specialists, the published versions include an extended introduction, as well as a foreword by the student's supervisor explaining the special relevance of the work for the field. As a whole, the series will provide a valuable resource both for newcomers to the research fields described, and for other scientists seeking detailed background information on special questions. Finally, it provides an accredited documentation of the valuable contributions made by today's younger generation of scientists.

Theses are accepted into the series by invited nomination only and must fulfill all of the following criteria

- They must be written in good English.
- The topic should fall within the confines of Chemistry, Physics, Earth Sciences, Engineering and related interdisciplinary fields such as Materials, Nanoscience, Chemical Engineering, Complex Systems and Biophysics.
- The work reported in the thesis must represent a significant scientific advance.
- If the thesis includes previously published material, permission to reproduce this must be gained from the respective copyright holder.
- They must have been examined and passed during the 12 months prior to nomination.
- Each thesis should include a foreword by the supervisor outlining the significance of its content.
- The theses should have a clearly defined structure including an introduction accessible to scientists not expert in that particular field.

Elahe Radmaneshfar

Mathematical Modelling of the Cell Cycle Stress Response

Doctoral Thesis accepted by
the University of Aberdeen, UK

Author	*Supervisor*
Dr Elahe Radmaneshfar	Dr Marco Thiel
Department of Physics	Department of Physics
Institute of Complex Systems and	Institute of Complex Systems and
Mathematical Biology	Mathematical Biology
University of Aberdeen	University of Aberdeen
Old Aberdeen	Old Aberdeen
UK	UK

ISSN 2190-5053   ISSN 2190-5061   (electronic)
ISBN 978-3-319-00743-4   ISBN 978-3-319-00744-1   (eBook)
DOI 10.1007/978-3-319-00744-1
Springer Cham Heidelberg New York Dordrecht London

Library of Congress Control Number: 2013943326

© Springer International Publishing Switzerland 2014

This work is subject to copyright. All rights are reserved by the Publisher, whether the whole or part of the material is concerned, specifically the rights of translation, reprinting, reuse of illustrations, recitation, broadcasting, reproduction on microfilms or in any other physical way, and transmission or information storage and retrieval, electronic adaptation, computer software, or by similar or dissimilar methodology now known or hereafter developed. Exempted from this legal reservation are brief excerpts in connection with reviews or scholarly analysis or material supplied specifically for the purpose of being entered and executed on a computer system, for exclusive use by the purchaser of the work. Duplication of this publication or parts thereof is permitted only under the provisions of the Copyright Law of the Publisher's location, in its current version, and permission for use must always be obtained from Springer. Permissions for use may be obtained through RightsLink at the Copyright Clearance Center. Violations are liable to prosecution under the respective Copyright Law. The use of general descriptive names, registered names, trademarks, service marks, etc. in this publication does not imply, even in the absence of a specific statement, that such names are exempt from the relevant protective laws and regulations and therefore free for general use.

While the advice and information in this book are believed to be true and accurate at the date of publication, neither the authors nor the editors nor the publisher can accept any legal responsibility for any errors or omissions that may be made. The publisher makes no warranty, express or implied, with respect to the material contained herein.

Printed on acid-free paper

Springer is part of Springer Science+Business Media (www.springer.com)

To my best friend: Mohammad

Supervisor's Foreword

This monograph introduces a comprehensive mathematical model for the cell cycle of *S. cerevisiae* and its responses to cellular stresses. The main evolutionary objective of every life form is to reproduce (or replicate) efficiently and, of course, survive long enough to do so. This determines what is often called the fitness of a species and ultimately its survival on evolutionary time scales. In a seminal paper, the mathematician and polymath John von Neumann described what he called "self-reproducing automata" and what is now known as "self-replicating machines". Roughly speaking, he examined how information can be stored and processed, and how an automaton can use this to self-replicate. One crucial component in computers to allow computations is the CPU clock-time. It allows coordinating (sub-) processes and orchestrating their interactions to successfully complete the objective of an algorithm.

To some extent, the cell cycle can be likened to the CPU clock-time generator as it allows biological cells to organise their survival and reproduction. In fact, the cell cycle is crucial for all life. A dysfunctional cell cycle can lead to cell death or, in some cases, to an uncontrolled cell division and ultimately to cancer. The cell cycle, in contrast to the CPU clock-time generator, also contributes to the information processing itself, and interacts with environmental cues. In fact, it is intricately linked to cellular pathways that react to a changing environment. These interactions are, which is typical in biological systems, highly complex.

Dr Radmaneshfar chooses a mathematical approach to condense the currently available biological knowledge into a predictive model. This mathematical description now allows us to better understand how different components of the cell cycle interact with stress response pathways; it helps to bring more order to the hitherto often somewhat disorganised collection of biological facts. It makes the processes and in particular their interdependence much clearer and makes the cross-talk between different cellular components accessible. Based on this mathematical description we can now better understand some counter-intuitive effects such as the behaviour of so-called MEN (Mitotic Exit Network) mutants, i.e., cell lines with a "disabled" exit from mitosis, a crucial step in cell division. Due to this defect, the cells cannot divide. Curiously, experiments have shown that if such cells are subjected to stresses, e.g. osmotic stress, they can successfully complete their cell division. This is a rather remarkable result, because it appears that first a central component of the cell's reproductive machinery is knocked out, and then to add fuel to the fire the cell is subjected to yet another attack. But in this particular situation the combination of these two wrongs, in fact, makes it a right. The cell's unexpected behaviour can now be understood, because the model highlights how the interaction of a vast number of components leads to a cascade of reactions that ultimately explain the underlying mechanism.

Apart from providing a comprehensive and very useful mathematical model, the work shows that the combination of the large and ever-growing body of biological knowledge with mathematical modelling can shed light on biological behaviour that cannot be understood by any one of the two disciplines alone. The history of science shows that the mathematisation of branches of study can greatly advance our knowledge. Areas such as physics, chemistry and computing science have all gained by adding mathematics to the portfolio of tools in the respective field. In many of these classical areas, scientists considered subsystems in isolation to facilitate the mathematical description. This compartmentalisation appears to be rather problematic in biological systems where the subsystems are intricately linked; in fact, so intricately that any modularisation is impossible. New mathematical approaches, like network theory, and recent advances in dynamical systems theory are making it possible now to describe complex biological systems to such a level of detail that the resulting models are not merely conceptual, but rather predictive. They can now help to elucidate biological mechanisms and make predictions that can be very relevant for the design of experimental studies.

The model presented in this monograph is a stepping stone towards a comprehensive mathematical description of a living cell; an objective that is still far from being achievable at our current state of knowledge. Enormous progress is currently being made in biology. This science has turned into one of the most exciting areas, if not the most exciting area, of all sciences. The new discoveries will profoundly change our lives and our perception of the world around us. Models like one presented here pave the way towards a new era in biology, which will be closely interwoven with mathematics. This will then mirror the development in other sciences such as physics and chemistry.

I hope that the reader will enjoy reading the book and the ideas therein. The book can serve as a guide into some of the most exciting questions in modern science.

Aberdeen, May 2013 Marco Thiel

Abstract

In this work, we investigate the response of a cell to different environmental conditions by developing two complementary models. The cell cycle is a sequence of complex biochemical events, which leads to replication and cell division. Moreover, cells are living in highly dynamic environments and exposed to various external and internal stresses. Hence, the cell cycle is tightly regulated to accomplish an error-free replication. The cell cycle control machinery does not operate in isolation and its network is intertwined with the elaborated stress signalling pathways. A systems biology approach is necessary to decipher the dynamics of this sophisticated processes. We introduce two different integrative mathematical models that, for the first time, describe the effect of osmotic stress and the α-factor on the progression through all stages of the cell cycle.

The first integrative model is based on ordinary differential equations. This model incorporates all known and several recently proposed interactions between the osmotic stress response pathway and the cell cycle. We study the response of the entire cell cycle to various levels of single-step osmotic stress. The key predictions of the model are: (i) exposure of cells to osmotic stress during the late S and the early G2/M phase can induce DNA re-replication before cell division occurs, (ii) cells stressed at the late G2/M phase display accelerated exit from mitosis and arrest in the next cell cycle, (iii) osmotic stress delays the G1-to-S and G2-to-M transitions in a dose dependent manner, whereas it accelerates the M-to-G1 transition independently of the stress dose and (iv) the Hog MAPK network compensates the role of the MEN network during cell division of MEN mutant cells. These model predictions are supported by experiments in *S. cerevisiae* and, moreover, have recently been observed in other eukaryotes. Furthermore, the model unveils the mechanisms that emerge as a consequence of the interaction between the cell cycle and stress response networks.

The second model is based on a discrete Boolean network. This model also includes all known interaction of osmotic stress and the α-factor with the cell cycle control network. The key results of this model are: (i) osmotic stress blocks the progression of the cell cycle in either of four possible arrest points, (ii) these arrests are all reversible, i.e. the cell can recover from the osmotic stress, (iii) the state of the cell at the onset of the stress dictates its final arrest point, (iv) a frequently used protocol to synchronise a population of cells based on exposure to the α-factor, is not a suitable technique to study one of the arrest points caused by osmotic stress and (v) osmotic stress can cause some of the *"non-dividing"* cells to divide.

In summary, these two models elucidate the physiological effect of the osmotic stress on the cell cycle and predict a series of novel interactions between the cell cycle and the osmotic stress response.

Aberdeen, May 2013   Elahe Radmaneshfar

Acknowledgments

Let me start my acknowledgments with my parents, to whom I owe where I am now. My mother lightened the passion for science in my spirit and my father taught me I can achieve whatever I aim for, thanks for giving me such important lessons that motivate me in all stages of my life since I can remember.

Over the past 3 years I have received support and encouragement from a great number of individuals. Dr M. Thiel has been a supervisor and a friend. His guidance, endless support and numerous fruitful discussions have illuminated this thoughtful and rewarding journey. I would like to thank Dr M. Carmen Romano for her support and countless hours she spent over the past few years as I moved from an idea to a completed study. This work has been done at the Institute for Complex Systems and Mathematical Biology. I would like to thank Prof. C. Grebogi for creating an inspiring scientific environment. I am very happy to be a part of this Institute and would like to thank all its members.

I have greatly appreciated the many enlightening discussions which I have had with my colleagues from CRISP project. In particular, I am grateful to (in alphabetical order) Prof. A. Brown, Prof. G. Coghill, Prof. N. Gow, Dr M. Jacobsen, Dr D. Kaloriti, Dr A. Moura, Dr W. Pang, Dr A. Tillmann, Dr C. Zabke, Dr T. You and Dr Z. Yin.

I am very thankful to my close cooperation partners Prof. M. C. Gustin (Rice University), Dr Y. Saka (University of Aberdeen) and Dr B. Schelter (University of Freiburg and University of Aberdeen), Dr A. Brand (University of Aberdeen) and Mr. D. Thomson (University of Aberdeen) for their patience and help in endless scientific discussions.

Especial thanks are due to my friends and colleagues from University of Aberdeen who help me to pass the critical moments: Dr L. Ciandrini, Dr E. Ullner, Dr A. Pokhilko, Mr. J. Karschau and Dr K. Chandrasekaran.

Many people I shared the office with in these three years: Mr. D. Clark, Ms. J. Slipantschuk, Ms. C. McLeman, Ms. H. Elsegai and many more. I had great time of learning and exchanging different cultures, thank you.

My experience of living in Scotland would not be possible without the help of a good friend Mr. N. Khan, who knows all of good and special places here. My friends from outside of the University of Aberdeen who were there all the time I need them Mrs. M. Bashiri, Dr E. Momtaz, Mrs. G. Mahjubian, Mr. M. Adib,

Dr E. Ebrahimi, Dr K. Nazarpour, Mr. M. Hallajin and Mrs. S. Nesyanpour and all my friends back at home who inspire my life despite all of the distances.

I am very thankful to my sisters (Samira, Shiva and Shirin) and my brothers (Saeed and Shahram) for their loving support in more wonderful ways than a page can hold.

The best of my acknowledgements goes to the most important and valuable friend of my life, my husband, Mohammad (M. Namavar), for his love, his inspiring persistence and patience with me. For never getting complacent, for always being honest with his critics, support and encourage me to improve. I am so lucky to share my experience of life with you. I am sure that this journey was not and is not possible without you! Thanks for being here and the way you are.

Contents

1 Introduction 1
  1.1 The Cell is a Complex Dynamical System 1
  1.2 Cell Cycle: A System Biology Perspective 2
  1.3 Stress Response 3
  1.4 Cell Cycle Response to Stresses 4
  References 6

2 A Biological Overview of the Cell Cycle and its Response to Osmotic Stress and the α-Factor 9
  2.1 Cell Cycle in Eukaryotes 9
    2.1.1 Cell Cycle Phases 9
    2.1.2 DNA Replication 10
    2.1.3 Mitosis and Cytokinesis 11
    2.1.4 Checkpoints 11
    2.1.5 Principles of the Cell Cycle Oscillation 12
  2.2 The Cell Cycle of *S. Cerevisiae* 13
    2.2.1 Principles of the Budding Yeast Cell Division 13
    2.2.2 The Molecular Mechanisms of the G1-to-S Transition ... 13
    2.2.3 The Molecular Mechanism of DNA Replication 14
    2.2.4 The Molecular Mechanism of the G2-to-M Transition ... 15
    2.2.5 The Molecular Mechanism of the M-to-G1 Transition ... 16
  2.3 Osmotic Stress Response 18
  2.4 Interaction Between the Cell Cycle Network and the Osmotic Stress Pathway 19
    2.4.1 Osmotic Stress Blocks the G1-to-S Transition 19
    2.4.2 Osmotic Stress Blocks the G2-to-M Transition 20
  2.5 Mating-Pheromone Response 22
    2.5.1 Pheromone Arrests the Cell Before START 23
  References 23

3 ODE Model of the Cell Cycle Response to Osmotic Stress 27
  3.1 Introduction 27
  3.2 Modelling Procedure and Assumptions 28

xiii

	3.2.1	Steps to Construct the Model	28
	3.2.2	Hypothesised Mechanisms in the Model	30
	3.2.3	Mathematical Definition of the Cell Cycle Phases	32
3.3	Model Description		32
	3.3.1	The Morphogenesis Checkpoint	32
	3.3.2	Regulation of Swe1	34
	3.3.3	Regulation of Hsl1-Hsl7 Complex in the Presence of Osmotic Stress	36
	3.3.4	Regulation of Mih1	37
	3.3.5	Regulation of Cdc28-Clb2 in the Presence of Osmotic Stress	37
	3.3.6	Cell Growth	41
	3.3.7	Influence of Hog1PP on the Cyclins Transcription	41
	3.3.8	Regulation of Sic1 Under Osmotic Stress	42
3.4	Parameter Estimation		45
3.5	Model Predictions		46
	3.5.1	Osmotic Stress Delays the G1-to-S and G2-to-M Transitions	47
	3.5.2	Osmotic Stress Causes Accelerated Exit from Mitosis	53
	3.5.3	Delays in the G1-to-S and G2-to-M Transitions are Dose Dependent, Whereas Acceleration of the M-to-G1 Transition is Dose Independent	56
	3.5.4	Osmotic Stress at Late S or Early G2/M Phase Causes DNA Re-Replication	56
	3.5.5	Stabilisation of Sic1 by Hog1PP Drives the Mitotic Exit in MEN Mutant Cells	58
3.6	Sensitivity Analysis of the Model		61
3.7	Model Validation in Laboratory		64
3.8	Discussion		64
	3.8.1	The Predictions of the Model are Supported by Various Biological Observations	65
	3.8.2	The Model Revealed Mechanisms for the Response of the Cell to Osmotic Stress	65
	3.8.3	The Relevance of the Model Predictions for Other Eukaryotes	66
References			67

4   Boolean Model of the Cell Cycle Response to Stress 71

4.1	Introduction		71
4.2	A Discrete Dynamical Model		73
4.3	State Transition Space of the Cell Cycle		75
	4.3.1	Deriving the State Transition Matrix	76
	4.3.2	Dynamical Properties of the Cell Cycle State Transition Matrix	77

		4.3.3	Basin of Attraction	78
	4.4	Results		78
		4.4.1	Cell Cycle State Transition	78
		4.4.2	Osmotic Stress Drives the Cell into One of the Four Fixed Points	80
		4.4.3	Biological Relevance of the Size of Basins	82
		4.4.4	Influence of the α-Factor Synchronisation on the Cell Cycle Dynamics	82
		4.4.5	Osmotic Stress Can Retrieve Some Frozen States to the Cell Cycle Trajectory	83
	4.5	Discussion		85
	References			86

5 Conclusion ... 89

Appendix A: List of Equations, Parameters and Initial Conditions ... 93

Appendix B: Effect of Methods of Update on Existence of Fixed Points ... 107

Chapter 1
Introduction

Cells as the fundamental unit of life perform a repetitive cycle of growth and division to reproduce. The sequence of biochemical events from "birth" to proliferation is called the *cell cycle*. This process is tightly regulated to allow cells to react accordingly to changes in their environment. One of the unique properties of cellular organisms is their ability to detect and respond selectively to extracellular signals – extracellular signal is called stress in this monograph. It is vital for a cell to react and adapt to stresses. This becomes apparent in many cases, e.g., under osmotic stress; "...*external osmolarity that is higher than the physiological range, can be a matter of life or death for all cells.*" [39].

In this work we study the impact of environmental changes on the cell cycle regulation from a molecular systems biology perspective. The core of cell cycle machinery and stress response are largely conserved in eukaryotes. Therefore, we focus on the model organism *S. cerevisiae* to elucidate the cell cycle responses to osmotic stress and the α-factor. We employ two different mathematical approaches. Our models unveil new aspects of cell cycle regulation which emerge as communication with external signals. In the next few pages, we present the challenges and open questions that inspired this research.

1.1 The Cell is a Complex Dynamical System

Defects in cell cycle is the cause of many human health problems. Therefore, molecular mechanisms that control the regulation of the cell cycle under different internal and external conditions keep fascinating many scientists over years. The breakthrough was the discovery of the "*universal*" control mechanisms of the cell cycle of eukaryotes in the late twentieth century [18, 21, 36]. Since then, due to the fast technological advances, a flood of new information about bewildering network of interactions has become available for different organisms. Hence, our detailed knowledge of the complex cell cycle and stress signalling network of various species is increasing

E. Radmaneshfar, *Mathematical Modelling of the Cell Cycle Stress Response*, Springer Theses, DOI: 10.1007/978-3-319-00744-1_1, © Springer International Publishing Switzerland 2014

rapidly. It is clear today that the cell cycle process and its responses to environmental conditions arise from intertwined nonlinear and dynamic interactions among large numbers of components [40, 48]. Yet, a clear understanding of how these pieces fit together into a coherent whole is still one of the big challenges of molecular biology. Investigation of the dynamical properties of these complex network is impossible by intuitive reasoning alone and requires a non-reductionist approach [32, 40].

Therefore, a systematic level of thinking is needed. As Nurse mentioned "...*Perhaps a proper understanding of the complex regulatory networks making up cellular systems like the cell cycle will require a similar shift from common sense thinking...*" [35]. Hence, the new approach to molecular biology is integrative, compared to resolute reductionism of the last century. In this framework, which is called "*systems biology*", pieces are put back together rather than taken apart [32, 48]. The field of systems biology aims to find general principles of living systems by amalgamating the computational approaches with experimental methods, and considering the network of cellular interactions rather than a simple constituent [25, 32].

Next, we review the advances in understanding the regulations of the cell cycle in various environmental conditions using systems biology approaches.

1.2 Cell Cycle: A System Biology Perspective

The first attempt to build a mathematical model for the cell cycle was performed to obtain a universal model that can explain the principle of its oscillation [16]. Later, Goldbeter developed a small model based on three differential equations for fission yeast. He developed this model to investigate the role of a particular negative feedback in deriving the cell cycle oscillation [17]. Tyson in the same year built a slightly more detailed model. That model predicted three modes of operation for the cell cycle: (i) spontaneous oscillation (stable limit cycle), (ii) stable steady state, which correspond to the so-called "checkpoints" in the cell cycle (see Sect. 2.1.4) and (iii) an excitable switch. Following the availability of more detailed data, the mathematical models of the cell cycle became more comprehensive [3, 5–7]. The most popular mathematical framework for modelling cell cycle is ordinary differential equations. Since molecular diffusion, transcription, translation and membrane transport are fast (a matter of seconds) compared with the duration of the cell cycle (hours), it is appropriate to use ordinary differential equations (ODEs) [48]. Furthermore, spatial localisation of reactions can be handled by compartmental modelling in this framework [48]. Other approaches, such as Boolean networks [10, 14, 28, 30], delay differential equations [44], and stochastic modelling [37, 51] have been successfully applied in deriving a mathematical description of the cell cycle of different organisms. The developed models are not only descriptive but also predictive. A successful example is the mathematical model of the budding yeast which was developed by Chen et al. based on the integration of pieces of knowledge from many experimental studies [6]. Some of the predictions of that model had been later validated by Cross et al. [9].

1.2 Cell Cycle: A System Biology Perspective

One of the great application of the theory of nonlinear dynamical systems is in explaining the dynamics of the events in the cell cycle [15, 48]. We now know that progression through the mitotic cell cycle has four crucial characteristics [50]; (i) The cell cycle is the sequence of events which are unidirectional and irreversible [19, 22, 29, 34, 41]. (ii) DNA replication and cell division are coordinated with the synthesis of all other cellular components (proteins, lipids, organelles, etc.). (iii) The cell cycle is an adaptive periodic process, i.e. the duration of each stage of the cell cycle is not fixed and is varied based on the successful completion of the essential tasks of that stage [48]. Checkpoints enforce the completion-requirements of tasks and block major transition of the cell cycle [12, 33, 47]. (iv) The molecular mechanisms that control all events of the cell cycle are extremely robust [10, 28, 30]. They perform a perfect DNA replication and cell division under a wide variety of conditions and stresses.

Cells receive and process a broad range of signals such as hormones, pheromones, temperature and osmotic pressure. The external signals are transduced to the cell by a cascade of components which is called signalling pathway. In this work our focus is on the influence of external osmotic pressure on the cell cycle of S. cerevisiae. Hence, we briefly review the advances that systems biology brings to our understanding of the signalling pathway and in particular osmotic stress signalling pathway.

1.3 Stress Response

Networks of signalling cascades are highly complex and include many cross-talks[1] [26]. Conserved building blocks of signalling pathways in eukaryotes include receptors, ERK (Extracellular signal-Regulated Kinases) or MAPK (Mitogen-activated protein kinases) cascades, G-proteins and small G-proteins [26].

To understand the complex behaviour of signalling networks, scientists have developed various models, ranging from universal models with the emphasise on some special features of signalling pathways [20], to comprehensive models that describe the dynamics of specific pathways in specific organisms [24, 27, 31, 52, 53]. Two different groups of systems biologists are interested in studying the signalling pathways.

The first group investigate the architecture, dynamics, and regulation of the signalling cascade [20, 23, 24, 31, 49]. These studies reveal the general design principle of the signalling pathway; (i) a signalling pathway can amplify its input signal [20], and (ii) it can have several different feedback loops, which operate on different time scale and are responsible mechanisms for fast response to the stress as well as adaptation to new environmental conditions [31, 49]. Furthermore, Kholodenko et al. showed that some of the negative feedback loops with a certain range of kinetic activity can cause oscillations [23]. In contrast, positive feedback loops result in

---

[1] referring to the case that two inputs work through distinct signalling pathways but cooperate to regulate the response [43].

4   1 Introduction

bistability, thereby ensuring that a weak stimulus transiently activates the pathway, while a strong stimulus results in a sustained activation [26].

The ultimate goal of the second group of system biologists is to develop an automated framework which can extrapolate the structure of a signalling pathway from the data set, such as DNA microarrays[2] [11, 42, 45, 46]. To this group, among all signalling pathways, the osmotic stress signalling pathway of yeast *S. cerevisiae* has been an intense subject of interest for various reasons: (i) osmotic stress is a common stress for budding yeast and can cause death, (ii) it is a nonlinear complex pathway, (iii) it is conserved in higher eukaryotes including humans and (iv) vast amount of quantitative and qualitative data is available and hence it is a rich workbench to test the algorithms which attempt to describe and investigate more complex signal transduction networks in higher eukaryotes [39].

Signalling pathways however do not work in isolation but govern a complex adaptive program that includes temporary arrest of cell cycle progression, and adjustment of transcription and translation patterns. Therefore, we give an overview of the advances in our understanding of the impact of signalling pathways activity on the cell cycle regulation from a computational biology point of view.

1.4 Cell Cycle Response to Stresses

The focus of this work is on the influence of osmotic stress on the cell cycle regulation of budding yeast. Since both the cell cycle control mechanism and signalling pathways are conserved in higher eukaryotes, this study will lead to elucidate the universal response of the cell cycle to stresses.

The cell responds to external signals and coordinates the regulation of the cell cycle to environmental conditions. The presence of osmotic stress activates a certain MAP kinase cascade which affects the cell cycle regulation. Several interactions between the osmotic stress signalling pathway and the control network of the cell cycle have recently been identified experimentally [2, 4, 8, 13]. It is known that osmotic stress can block the cell cycle progression at two different stages, which are located before two main cell cycle events; namely DNA replication and cell division [2, 4, 8, 13]. However, despite this experimental observation many more aspects of the cell cycle response to osmotic stress are still unknown, for example: the impact of the stress on the regulation of two main cell cycle events (DNA replication and cell division); the dependency of the cell cycle response to the various levels of the stress; the dynamics of the cell cycle regulation in the presence of the stress; the activation of new checkpoints due to the presence of the stress; the response of various mutated cells to osmotic stress.

The complexity of the cell cycle control network, the osmotic stress signalling pathway, and their interaction require systems biology approaches to answer these

---

[2] A DNA microarray (also known as DNA chip or biochip) is commonly used to measure the expression levels of large numbers of genes concurrently.

1.4 Cell Cycle Response to Stresses

open questions. Hitherto the only existing mathematical model (for the response of the budding yeast cell cycle to osmotic stress) was developed to study the the impact of the osmotic stress signalling pathway on only a single specific stage of the cell cycle [1]. However, the stress signalling network does not impact on just one isolated stage of the cell as assumed in the model of Adrover et al. [1]. Moreover, this model does not discuss the influence of osmotic stress on DNA replication and cell division.

Environmental stresses can occur at any stage of the cell cycle. Importantly, the cell cycle stages are intertwined. Therefore to study the regulation of all cell cycle events, which are controlled by a complex and interconnected network, in the presence of osmotic stress, we develop two novel integrative mathematical models that predict the influence of osmotic stress on all stages of the cell cycle for the first time. These models are based on two different computational frameworks; continuous and discrete.

In the first modelling approach we incorporate the osmotic stress response network with all stages of the cell cycle machinery of budding yeast, *S. cerevisiae*, using ordinary differential equations (ODEs). We develop this model bearing in mind that: *"successful integration at the systems level must be built on successful reduction, but reduction alone is far from sufficient."* [32]. This integrative mathematical model provides a series of novel predictions and it also elucidates how the elaborate cell cycle works in the presence of osmotic stress. Our model predicts the influence of various levels of osmotic stress on DNA replication and cell division. It indicates the existence of new checkpoint in response to osmotic stress. Moreover, it investigates the molecular processes and the cell dynamics under frequently changing environmental conditions. Furthermore, the model explains the causes of the recent experimental observations.

The second modelling approach, using discrete Boolean networks, brings the dynamical changes of the cell cycle in response to osmotic stress and the α-factor to light. Since this approach leads to a finite number of possible states, it reveals the state transitions of the cell cycle and its adaptation in various conditions of growth.

In the next four chapters we present the modelling approaches and their predictions. This provides more insights into cell cycle control mechanisms under various environmental conditions.

A concise introduction about the cell cycle control mechanisms which are conserved among species is given in Chap. 2. To construct a practical model, a proper perception of the regulatory mechanisms of the system is needed. Hence, in this chapter we review the known biological mechanisms of (i) the cell cycle machinery of *S. cerevisiae*, and (ii) signalling networks of osmotic stress and the α-factor. Furthermore, we explore the known coupling and the interactions of the osmotic stress and the α-factor to the cell cycle. We use these mechanisms to establish the two models.

The ODE model is presented in Chap. 3. We explain step-by-step our approach in deriving the model including parameterisation, model validation and model predictions. We study the influence of different levels of osmotic stress at various stages of the cell cycle. This model does not only explain the known biological mechanisms but also elucidates the molecular interactions of the cell cycle of *S. cerevisiae* which emerge in response to osmotic stress. Furthermore, it characterises the role of the cell cycle components in these interactions. It also predicts responses of cells to the osmotic stress, according to the state of the cells at the time of activation of the stress response. We also explain the experiments that are being conducted by our experimental collaborators to validate the prediction of the models.

The second model – which is based on discrete mathematics and Boolean networks – is introduced in Chap. 4. This model explores the dynamical behaviour of budding yeast in response to the osmotic stress and the α-factor. Furthermore, we explain the translation of the Boolean model to the state transition space. We analyse the dynamical properties of this space under different environmental conditions to reveal the influence of osmotic stress and the α-factor on the dynamics of the cell cycle.

Finally, in Chap. 5, we summarise the conclusions and bring together the different facets of the cell cycle response to environmental stresses. Moreover, we briefly present the work in progress regarding the cell cycle response to a fluctuating environment.

References

1. M.A. Adrover, Z. Zi, A. Duch, J. Schaber, A. Gonzalez-Novo, J. Jimenez, M. Nadal-Ribelles, J. Clotet, E. Klipp, F. Posas. Time-dependent quantitative multicomponent control of the G1-S network by the stress-activated protein kinase Hog1 upon osmostress. Sci. Signal. **4**(192), 63 (2011)
2. M.R. Alexander, M. Tyers, M. Perret, B.M. Craig, K.S. Fang, M.C. Gustin, Regulation of cell cycle progression by Swe1p and Hog1p following hypertonic stress. Mol. Biol. Cell **12**(1), 53–62 (2001)
3. M. Barberis, E. Klipp, M. Vanoni, L. Alberghina, Cell size at S phase initiation: an emergent property of the G1/S network. PLoS Comput. Biol. **3**(4), 649–666 (2007)
4. G. Bellí, E. Garí, M. Aldea, E. Herrero, Osmotic stress causes a G1 cell cycle delay and downregulation of Cln3/Cdc28 activity in *Saccharomyces cerevisiae*. Mol. Microbiol. **39**(4), 1022–1035 (2001)
5. G. Charvin, C. Oikonomou, E.D. Siggia, F.R. Cross. Origin of irreversibility of cell cycle start in budding yeast. PLoS Biol. **8**(1), e1000284 (2010)
6. K.C. Chen, A. Csikasz-Nagy, B. Gyorffy, J. Val, B. Novak, J.J. Tyson, Kinetic analysis of a molecular model of the budding yeast cell cycle. Mol. Biol. Cell. **11**(1), 369–391 (2000)
7. K.C. Chen, L. Calzone, A. Csikasz-nagy, F.R. Cross, B. Novak, J.J. Tyson, Integrative analysis of cell cycle control inbBudding yeast. Mol. Biol. Cell. **15**, 3841–3862 (2004)
8. J. Clotet, X. Escoté, M.A. Adrover, G. Yaakov, E. Garí, M. Aldea, E. de Nadal, F. Posas, Phosphorylation of Hsl1 by Hog1 leads to a G2 arrest essential for cell survival at high osmolarity. EMBO J. **25**(11), 2338–2346 (2006)
9. F.R. Cross, V. Archambault, M. Miller, M. Klovstad, Testing a mathematical model of the yeast cell cycle. Mol. Biol. Cell. **13**(1), 52–70 (2002)
10. M.I. Davidich, S. Bornholdt. Boolean network model predicts cell cycle sequence of fission yeast. PLoS ONE **3**(2), e1672 (2008)
11. P. Diner, J. Veide Vilg, J. Kjellen, I. Migdal, T. Andersson, M. Gebbia, G. Giaever, C. Nislow, S. Hohmann, R. Wysocki, M.J. Tamas, M. Grotli. Design, synthesis, and characterization of a highly effective Hog1 inhibitor: a powerful tool for analyzing map kinase signaling in yeast. PLoS ONE **6**(5), e20012 (2011)

References

12. S.J. Elledge, Cell cycle checkpoints: preventing an identity crisis. Science **274**(5293), 1664–1672 (1996)
13. X. Escoté, M. Zapater, J. Clotet, F. Posas, Hog1 mediates cell-cycle arrest in G1 phase by the dual targeting of Sic1. Nat. Cell Biol. **6**(10), 997–1002 (2004)
14. A. Faure, A. Naldi, C. Chaouiya, D. Thieffry, Dynamical analysis of a generic Boolean model for the control of the mammalian cell cycle. Bioinformatics **22**(14), e124–e131 (2006)
15. J.E. Ferrell Jr, T.Y.C. Tsai, Q. Yang, Modeling the cell cycle: why do certain circuits oscillate? Cell **144**(6), 874–885 (2011)
16. D.A. Gilbert, The nature of the cell cycle and the control of cell proliferation. BioSystems **5**(4), 197–206 (1974)
17. A. Goldbeter, A minimal cascade model for the mitotic oscillator involving cyclin and cdc kinase. PNAS **88**(20), 9107–9111 (1991)
18. L.H. Hartwell, Nobel Lecture. Yeast and cancer. Biosci. Rep. **22**(3–4), 373–394 (2002)
19. E. He, O. Kapuy, R.A. Oliveira, F. Uhlmann, J.J. Tyson, B. Novak, System-level feedbacks make the anaphase switch irreversible. PNAS **108**(24), 10016–10021 (2011)
20. R. Heinrich, B.G. Neel, T.A. Rapoport, Mathematical models of protein kinase signal transduction. Mol. Cell. **9**(5), 957–970 (2002)
21. T. Hunt, Noble Lecture. Protein synthesis, proteolysis, and cell cycle transition. Biosci. Rep. **22**(5–6), 465–486 (2002)
22. O. Kapuy, E. He, S. Lopez-Aviles, F. Uhlmann, J.J. Tyson, B. Novak, System-level feedbacks control cell cycle progression. FEBS Lett. **583**(24), 3992–3998 (2009)
23. B.N. Kholodenko, Negative feedback and ultrasensitivity can bring about oscillations in the mitogen-activated protein kinase cascades. Eur. J. Biochem. **267**(6), 1583–1588 (2000)
24. E. Klipp, B. Nordlander, R. Kröger, P. Gennemark, S. Hohmann, Integrative model of the response of yeast to osmotic shock. Nat. Biotechnol. **23**(8), 975–982 (2005)
25. E. Klipp, R. Herwig, A. Kowald, C. Wierling, H. Lehrach. Systems biology in practice: concepts, implementation and application. Wiley VCH, (2006)
26. E. Klipp, W. Liebermeister. Mathematical modeling of intracellular signaling pathways. BMC Neurosci. **7** (2006)
27. B. Kofahl, E. Klipp, Modelling the dynamics of the yeast pheromone pathway. Yeast **21**(10), 831–850 (2004)
28. F. Li, T. Long, Y. Lu, Q. Ouyang, C. Tang, The yeast cell-cycle network is robustly designed. PNAS **101**(14), 4781–4786 (2004)
29. S. Lopez-Aviles, O. Kapuy, B. Novak, F. Uhlmann, Irreversibility of mitotic exit is the consequence of systems-level feedback. Nature **459**(7246), 592–595 (2009)
30. K. Mangla, D.L. Dill, M.A. Horowitz. Timing robustness in the budding and fission yeast cell cycles. PloS ONE **5**(2), e8906 (2010)
31. J.T. Mettetal, D. Muzzey, C. Gomez-Uribe, A. Van Oudenaarden, The frequency dependence of osmo-adaptation in *Saccharomyces cerevisiae*. Science **319**(5862), 482–484 (2008)
32. D. Noble, *The Music of Life: Biology Beyond the Genome* (Oxford University Press, Oxford, 2006)
33. B. Novak, J.J. Tyson, Quantitative analysis of a molecular model of mitotic control in fission yeast. J. Theor. Biol. **173**(3), 283–305 (1995)
34. B. Novak, J.J. Tyson, B. Gyorffy, A. Csikasz-Nagy, Irreversible cell-cycle transitions are due to systems-level feedback. Nat. Cell Biol. **9**(7), 724–728 (2007)
35. P.M. Nurse, A long twentieth century of the cell cycle and beyond. Cell **100**(1), 71–78 (2000)
36. P.M. Nurse, Nobel Lecture. Cyclin dependent kinases and cell cycle control. Biosci. Rep. **22**(5–6), 487–499 (2002)
37. Y. Okabe, M. Sasai, Stable stochastic dynamics in yeast cell cycle. Biophys. J. **93**(10), 3451–3459 (2007)
38. E. Radmaneshfar, M. Thiel, Recovery from stress: a cell cycle perspective. J. Comp. Int. Sci. **3**(1–2), 33–44 (2012)
39. H. Saito, F. Posas, Response to hyperosmotic stress. Genetics **192**(2), 289–318 (2012)

40. U. Sauer, M. Heinemann, N. Zamboni, Getting closer to the whole picture. Science **316**(5824), 550–551 (2007)
41. R.M. Scaife, Selective and irreversible cell cycle inhibition by diphenyleneiodonium. Mol. Cancer Ther. **4**(6), 876–884 (2005)
42. J. Schaber, M. Flottmann, J. Li, C.F. Tiger, S. Hohmann, E. Klipp, Automated ensemble modeling with modelMaGe: analyzing feedback mechanisms in the Sho1 branch of the HOG pathway. PLoS ONE **6**(3), e14791 (2011)
43. M.A. Schwartz, M.H. Ginsberg, Networks and crosstalk: integrin signalling spreads. Nat. Cell Biol. **4**(4), E65–E68 (2002)
44. J. Srividhya, M.S. Gopinathan, A simple time delay model for eukaryotic cell cycle. J. Theor. Biol. **241**(3), 617–627 (2006)
45. M. Steffen, A. Petti, J. Aach, P. Dhaeseleer, G. Church, Automated modelling of signal transduction networks. BMC Bioinformatics. 3 (2002)
46. C.F. Tiger, F. Krause, G. Cedersund, R. Palmer, E. Klipp, S. Hohmann, H. Kitano, M. Krantz, A framework for mapping, visualisation and automatic model creation of signal-transduction networks. Mol. Syst. Biol. **8** (2012)
47. J.J. Tyson, B. Novak, K. Chen, J. Val, Checkpoints in the cell cycle from a modeler's perspective. Prog. Cell Cycle Res. **1**, 1–8 (1995)
48. J.J. Tyson, K. Chen, B. Novak, Network dynamics and cell physiology. Nat. Rev. Mol. Cell Biol. **2**(12), 908–916 (2001)
49. J.J. Tyson, K.C. Chen, B. Novak, Sniffers, buzzers, toggles and blinkers: dynamics of regulatory and signaling pathways in the cell. Curr. Opin. Cell Biol. **15**(2), 221–231 (2003)
50. J.J. Tyson, B. Novak, Temporal organization of the cell cycle. Curr. Biol. **18**(17), R759–R768 (2008)
51. Y. Zhang, M. Qian, Q. Ouyang, M. Deng, F. Li, C. Tang, Stochastic model of yeast cell-cycle network. Physica D: Nonlinear Phenom. **219**(1), 35–39 (2006)
52. Z. Zi, W. Liebermeister, E. Klipp, A quantitative study of the Hog1 MAPK Response to fluctuating osmotic stress in *Saccharomyces cerevisiae*. PLoS ONE **5**(3), e9522 (2010)
53. X. Zou, T. Peng, Z. Pan, Modeling specificity in the yeast MAPK signaling networks. J. Theor. Biol. **250**(1), 139–155 (2008)

Chapter 2
A Biological Overview of the Cell Cycle and its Response to Osmotic Stress and the α-Factor

2.1 Cell Cycle in Eukaryotes

The word "cell" originates from Latin *cella*, which means "small room". It was applied for the first time by Hooke in his book *Micrographia* in September 1665. There are 10 to perhaps 100 million distinct life forms in the world [1, 32], many of which consist of various types of cells. Although different cells have disparate functions and shapes, they still have a similar basic chemistry [35].

Cells can be classified into (i) eukaryotic cells which contain a nucleus, and (ii) prokaryotic cells (such as bacteria) which lack any distinction between nucleus and cytoplasm [1, 32]. A cell can reproduce by growing and dividing into two daughter cells with the same genetic information [1, 32]. Single-celled organisms reproduce by cell division, whereas multicellular organisms need countless cell divisions to generate from founder cells (stem cells) diverse communities of cells that build the organs and tissues [1, 32].

Before each division all components and the "machinery" of a cell have to be duplicated, to allow the daughter cell to repeat the process. One crucial event of the cell cycle is the duplication of DNA to pass the genetic information to the next generation. Another crucial part of the cell cycle is the chromosome segregation, in which the sister chromatids are distributed to each daughter cell [1, 32]. The principle molecular machinery of the cell cycle that regulates DNA replication and chromosome segregation is preserved in eukaryotes.

2.1.1 Cell Cycle Phases

Cell cycle consists of four different phases: G1, S, G2 and M. In G1 phase cells increase in size and prepare for DNA replication. In S phase DNA replication occurs. From one double-stranded DNA molecule (chromosome) two identical double stranded DNA molecules (sister chromatids) are formed and held together

E. Radmaneshfar, *Mathematical Modelling of the Cell Cycle Stress Response*,
Springer Theses, DOI: 10.1007/978-3-319-00744-1_2,
© Springer International Publishing Switzerland 2014

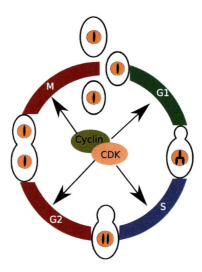

Fig. 2.1 Diagram of the eukaryotic cell cycle. G1 is the gap before DNA synthesis, during which the cell grows to a critical size. During S phase, DNA replication takes place. G2 phase is the gap between the S phase and the M phase. In G2 phase a set of control mechanism ensures that everything is ready to enter the M phase. Finally, during the M phase chromosome segregation takes place and the cell divides into two cells. G1, S and G2 phases are called *interphase*

with cohesin proteins. In G2 phase the accuracy of DNA replication and the morphology of the cell is checked. Finally, in M phase the cell divides into two daughter cells, each of which has a complete copy of the chromosome (see Fig. 2.1).

The details of the mechanism of the cell cycle are not identical in different types of cells [32]. Despite these differences, the main machinery of the cell cycle events – DNA replication; mitosis; cytokinesis and the presence of checkpoints – is highly conserved throughout all organisms.

2.1.2 DNA Replication

Before each cell division, a cell must duplicate its genome faithfully to avoid mutations. Also, DNA replication should happen once and only once per cell cycle. The synthesis and duplication of DNA is a complex process which fills a major fraction of the cell cycle. It starts in the late M phase and the early G1 phase, during which an initiator protein complex called pre-Replicative Complex (pre-RC), assembles at each origin of replication. This step is called licensing [1, 32]. The pre-RC complex consists of an Origin Recognition Complex (ORC), Mini-chromosome maintenance complex (Mcm) and licensing proteins. Once the pre-RC has mounted at the origins of replication in the late G1 phase, the origins are ready to fire [1, 32]. The S phase cyclins triggers the DNA synthesis at the onset of the S phase. The

S phase-dependent kinases (S-phase cyclins) are deactivated in the late M phase. The reformation of pre-RC is inhibited, while the S-phase cyclins are activated. This ensures the occurrence of the DNA replication only once per cell cycle [1, 32].

2.1.3 Mitosis and Cytokinesis

At the onset of the M phase, every cell has a tightly associated pair of chromosomes called sister chromatids. During the M phase, sister chromatids are segregated and distributed to each nascent cell [1, 32, 34]. This is followed by cytokinesis.

Mitosis is usually divided into four distinct stages in yeast: prophase, metaphase, anaphase and telophase. The prophase is the first stage during which chromosome condensation, centrosome separation and spindle assembly initiate. It is followed by the metaphase, during which the sisters are aligned at the centre of the spindle, anticipating the signal to separate in the anaphase. At the onset of the anaphase, the cohesin links between sister chromatids are abruptly removed and the separated sisters are dragged to the opposite poles of the spindle. Finally, the mitosis is completed in the telophase. During the telophase the chromosomes and other nuclear components are re-bundled into daughter nuclei [1, 32, 34]. Cytokinesis is the process of cell division, which happens at the end of mitosis. Its regulation depends on the progression through mitosis. The events of the mitosis are tightly controlled by the M phase cyclins. We will explain the control mechanism of the M phase for the model organism *Saccharomyces cerevisiae* in Sect. 2.1.5.

2.1.4 Checkpoints

There are several checkpoints in the cell cycle that control the cell cycle progression [25]. The main role of each checkpoint is to ensure whether certain conditions are satisfied before entering the next phase. If some irregularity is detected the cell is arrested until the problem is corrected. There are three main checkpoints operate at the end of G1, G2 and during M phase, respectively [18, 25]. These are example of conserved checkpoints among eukaryotes, but note that each species have several specific ones.

At the G1 checkpoints the size and also the DNA content of the cell are examined. If the cell is large enough and its DNA is intact it will proceed to the S phase. Otherwise it remains in the G1 phase [25]. At the G2 checkpoints the quality of the DNA synthesis, the size and the polarity of the cell are checked. If these conditions meet the required criteria, the cell will proceed to its M phase [25]. At the M phase checkpoint, which is referred to as metaphase checkpoint, the alignment of the chromosomes and the completeness of DNA replication are checked. Furthermore, the spindles need to be oriented towards the daughter cell. If all these conditions are met, the cell can divide [25].

2.1.5 Principles of the Cell Cycle Oscillation

The cell cycle is controlled by a complex molecular network which is responsible for the self-sustained oscillatory expression of a large set of genes. There are three main transitions in the cell cycle: the G1-to-S transition (START), the G2-to-M transition and the M-to-G1 transition (FINISH).

Cell cycle oscillations and transitions from one phase to the next are caused by the activation and inactivation of a complex that is composed of two main subunits: a catalytic subunit and a regulatory subunit (see Fig. 2.1). A catalytic subunit – Cyclin-dependent kinase (Cdk) – forms a dimer with a regulatory subunit cyclin [31]. These kinase complexes (Cdk-cyclins) control the cell cycle oscillation, as well as cell size and DNA replication. The activity of Cdk is regulated by the availability of its cyclin partners, binding to stoichiometric Cdk inhibitors, and inhibitory tyrosine phosphorylation [31].

The molecular machinery which regulates DNA replication and segregation is highly conserved, from unicellular eukaryotes, such as yeasts, to multicellular eukaryotes [35]. Therefore, simple eukaryotes, such as fission yeast and budding yeast, serve as model organisms to understand the cell cycle control mechanisms in Metazoa including humans. To understand such a complex system we have focused on the cell cycle machinery of budding yeast, *S. cerevisiae*, and its response to osmotic stress and the α-factor, and we have developed two mathematical models [39, 40], which are presented in Chaps. 3 and 4.

Our first model is based on Ordinary Differential Equation (ODE), which predicts the influence of osmotic stress in all stages of cell cycle. This model integrates the osmotic stress response network with the cell cycle machinery of budding yeast, *S. cerevisiae*. Since the cell cycle phases are interlinked, the effect of osmotic stress on the cell cycle cannot be predicted just by considering one single phase. We introduce a comprehensive mathematical model, which predicts how this elaborate system might work in the presence of osmotic stress. It also provides a workbench for further investigation of the molecular processes and the cell behaviour under various environmental conditions and experimental setups.

The second model is a Boolean representation of the cell cycle reactions to stresses, which investigates the dynamical properties of the cell cycle oscillation in the presence of multiple stresses; the osmotic stress and the α-factor. Since it is a discrete model, study of the entire finite set of the cell cycle states under various environmental conditions is achievable. This model shows that the cell cycle is robustly designed and it can recover from osmotic stress and the α-factor induced arrest.

The models are based on known interactions that control the yeast cell cycle and its response to osmotic stress and the α-factor, which we will review briefly in the next section.

2.2 The Cell Cycle of S. Cerevisiae

2.2.1 Principles of the Budding Yeast Cell Division

In budding yeast, S. cerevisiae, the cell cycle is controlled by a robust molecular network, which consists of more than 800 genes [46]. Note that the G2 phase in S. cerevisiae is usually very short compared to the M phase, often not distinguished from the M phase and referred to as the G2/M phase. S. cerevisiae replicates by budding asymmetrically [24]. Daughter cells are smaller than mother cells, and hence the G1 phase of a daughter cell is longer than that of the mother cell [24, 32].

S. cerevisiae has five Cdks (Cdc28, Pho85, Kin28, Ssn3 and Ctk1). Cdc28 is the central coordinator of the major cell cycle events [31, 52]. The function of the other Cdks has not fully been understood yet [31]. In budding yeast Cdc28 constitutively expressed. The activity of Cdc28 is regulated by the availability of its cyclin partners, inhibitory tyrosine phosphorylation and binding to stoichiometric Cdk inhibitors (Sic1, Cdc6). Cdc28 has two types of associated cyclins: (i) three G1 cyclins (Cln1, Cln2, Cln3) and (ii) six B-type cyclins (Clb1 to Clb6) [31].

G1 cyclins regulate the events in the gap between mitosis and DNA replication, whereas B-type cyclins are expressed successively from START to FINISH. Among the G1 cyclins, Cln3 is responsible for cell growth and the activation of Cln1 and Cln2 [52]. G1 cyclins are essential for cell cycle progression. The cell cycle is arrested if the cell loses all Clns. However, if at least one of the G1 cyclins is expressed, the cell will still proceed to START [41]. As Cln1 and Cln2 have redundant functions [52], we will refer to both of them as Cln2. The activity of Cdc28-Cln2 is maximal at START (see Fig. 2.2) [17]. $cln1\Delta cln2\Delta$ cells initiate DNA synthesis with a delay and show slow growth [22].

The six B-type cyclins are divided into three distinct pairs of similar functions [31]:

(i) The major functions of Clb5 and Clb6, both represented by Clb5 from now on, are the initiation of the DNA synthesis, and the negative regulation of Cdc28-Cln2 activity [6]. $clb5\Delta clb6\Delta$ cells exhibit long delay in initiating the S phase; but the length of S phase is comparable to wild-type cells [6].

(ii) The mitotic cyclins Clb1 and Clb2 [28] are represented by Clb2 from now on. Clb2 is crucial for successful mitosis and $clb1\Delta clb2\Delta$ cells are not viable [48].

(iii) The remaining B-cyclins, Clb3 and Clb4, have redundant roles in S phase initiation and spindle formation [31].

2.2.2 The Molecular Mechanisms of the G1-to-S Transition

The G1-to-S phase transition – START – refers to the transcriptional cascade that triggers three main events: budding, DNA replication, and the duplication of spindle pole bodies [17]. The Cdc28-Cln3 complex catalyses START; it activates the

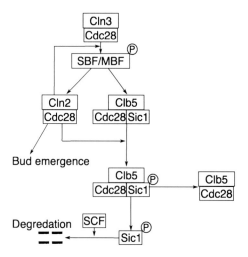

Fig. 2.2 G1-to-S transition: Cdc28-Cln3 complex activates the SBF and MBF transcription factors. These are transcription factors of *CLN2* and *CLB5* genes. Expression of *CLN2* and *CLB5* increases the levels of Cdc28-Cln2 and Cdc28-Clb5. The activation of Cdc28-Cln2 leads to the bud emergence. Moreover, there is a positive feedback loop from Cdc28-Cln2 to Cdc28-Cln3, which makes the G1-to-S transition irreversible. Sic1 maintains Cdc28-Clb5 in an inactive form during the G1 phase. To enable the G1-to-S transition, Sic1 needs to be phosphorylated and inactivated. Phosphorylated Sic1 is targeted for degradation by the SCF (Skp1-cullin-F-box) complex. The absence of Sic1 frees Cdc28-Clb5. Activity of Cdc28-Clb5 initiates DNA replication

transcription factor complexes: SBF and MBF (see Fig. 2.2). These, in turn, trigger the transcription of a set of more than 200 genes [27]; in particular *CLN2* and *CLB5*. The expression of *CLN2* and *CLB5* leads to the increased level of Cdc28-Cln2 and Cdc28-Clb5. The activation of Cdc28-Cln2 induces the bud formation. Moreover, there is a positive feedback loop from Cdc28-Cln2 to Cdc28-Cln3 (see Fig. 2.2), which makes this transition switch-like and irreversible [12, 17].

Sic1, the Cyclin Kinase Inhibitor (CKI) in the G1 phase, maintains Cdc28-Clb5 in an inactive form during most of START [42]. To enable the G1-to-S transition, Sic1 has to be inactivated. Cdc28 phosphorylates Sic1 [42] and targets it for degradation by the SCF (Skp1-Cullin-F-box) complex [33]. The absence of Sic1 triggers the activity of Cdc28-Clb5, which initiates the DNA replication and transfers the cell to the S phase (see Fig. 2.2).

2.2.3 The Molecular Mechanism of DNA Replication

DNA replication is a multi-step process which begins at the origins of replication (see Fig. 2.3). The multi-protein complex ORC (Origin Recognition Complex) remains bound to the origins of replication throughout the cell cycle. The first step of DNA

2.2 The Cell Cycle of S. Cerevisiae

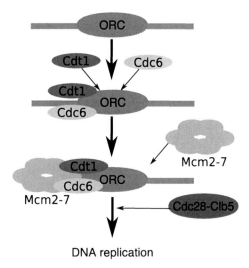

Fig. 2.3 DNA replication is a multi-step process which begins at the origins of replication. The multi-protein complex ORC remains bound to the origins of replication throughout the cell cycle. In the first step, the pre-replicative complex consisting of Cdc6, Cdt1 and Mcm complex binds to ORC. In the second step, DNA replication is triggered by Cdc28-Clb5

replication takes place during the late M phase and the early G1 phase. In this step, the pre-replicative complex, required for the initiation of DNA replication [16], binds to the ORC. The pre-replicative complex consists of the proteins Cdc6, Cdt1 and a complex of six proteins called Mcm (Mini-chromosome maintenance).

The second step occurs at the onset of the S phase (see Fig. 2.3). Activity of Cdc28-Clb5 first triggers DNA replication and later on disassembles the pre-replicative complex from the origins of replication. It also phosphorylates Cdc6, thereby promotes Cdc6 degradation and prevents *de novo* assembly of the pre-replicative complex during the G2/M phase. Therefore, Cdc28-Clb5 has a dual role: it enables the initiation of DNA replication; it also warrants that DNA replication happens once and only once before cell division [15].

2.2.4 The Molecular Mechanism of the G2-to-M Transition

The next cell cycle transition, G2-to-M, is mainly governed by the activity of Cdc28-Clb2 [48], which in turn is regulated by several mechanisms:

(i) The protein kinase Swe1 inhibits Cdc28-Clb2 activity by tyrosine phosphorylation of Cdc28 [43]. Swe1 is part of the morphogenesis checkpoint machinery, and it prevents the cell from entering into the G2 phase when aspects of bud formation are defective [43, 44]. Accumulation of Swe1 begins in the S phase [4] and it degrades quickly, when it is hyper-phosphorylated by the Hsl1-Hsl7 complex and

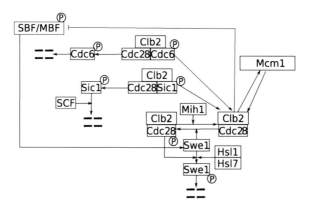

Fig. 2.4 G2-to-M transition: the activity of Cdc28-Clb2 controls the G2-to-M transition. The activity of this cyclin-Cdk complex is controlled via several mechanisms. Tyrosine phosphorylation of Cdc28 that is mediated by Swe1 inactivates Cdc28-Clb2. Moreover, Clb2 is transcriptionally controlled by Mcm1, which is activated by Cdc28-Clb2, thereby establishing a positive feedback loop. Active Cdc28-Clb2 also inactivates SBF/MBF transcription factors and promotes the G2-to-M transition

Cdc28-Clb2. It is worth mentioning that the phosphatase Mih1 reverses the tyrosine phosphorylation of Cdc28 [43].

(ii) The expression of Clb2 is transcriptionally controlled. Cdc28-Clb2 activates Mcm1 [3], which in turn promotes the transcription of *CLB2*, thereby establishing a positive feedback loop (see Fig. 2.4). Active Cdc28-Clb2, therefore, mediates the G2-to-M transition, during which the replicated chromosomes are segregated and nuclear division takes place (see Fig. 2.6).

(iii) The degradation of Clb2 is required for the M-to-G1 transition. We will discuss this transition in the next section.

2.2.5 The Molecular Mechanism of the M-to-G1 Transition

The M-to-G1 transition – the exit from mitosis – is achieved mainly by the inactivation of Cdc28-Clb2 [47]. Cdc20 and Cdh1, two major APC (Anaphase Promoting Complex) components, inactivate Cdc28-Clb2 [7, 53]. First, Mcm1 mediates the activation of the Cdc20 during the anaphase; then, Cdc20 triggers the pathway that results in the activation of Cdh1 and Cdc14 [7, 47, 53]. Cdh1 degrades the remaining fraction of Clb2 during the exit from mitosis (see Fig. 2.5) [7]. Finally, the phosphatase Cdc14 promotes the transition to the G1 phase in two steps: (i) it activates Swi5, a transcription factor of Sic1 and Cdc6 [50]; (ii) it dephosphorylates Cdh1, Sic1 and Cdc6. Active Sic1 and Cdc6 bind directly to Cdc28-Clb2, resulting in its inactivation (see Fig. 2.5).

2.2 The Cell Cycle of S. Cerevisiae

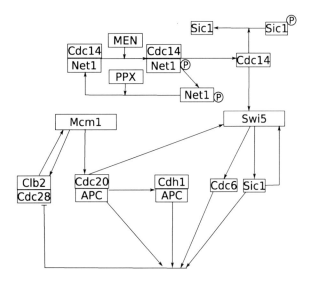

Fig. 2.5 M-to-G1 transition: this transition is achieved by the inactivation of Cdc28-Clb2. Cdc20 primarily degrades Clb2 and also mediates activation of Cdh1 and Cdc14. Then Cdh1 degrades the remaining fraction of Clb2 during the exit from mitosis. Cdc14 stimulates the activity of Sic1, Cdc6 and Cdh1 and therefore mediates the M-to-G1 transition

Figure 2.6 summarises the molecular mechanisms that control the cell cycle of *S. cerevisiae*. As explained above, the progression through the cell cycle occurs via constitutive activation of the corresponding cyclins in different cell cycle phases. The cell needs to pass through three main transitions successfully before dividing into two daughter cells. This succession of biochemical events comprising the cell cycle is distorted in the presence of stress. In order to study the response of the cell to different stresses it is important to understand: (i) the stress response signalling pathway, and (ii) the coupling of this pathway to the cell cycle control network. We have studied the cell cycle of *S. cerevisiae* under two different external signals; osmotic stress and the mating pheromone (α-factor). These signalling networks have been thoroughly studied in the past and found to affect the cell cycle progression. It has recently become clear that to study the response of the cell cycle to stress, the cell cycle and the stress signalling network have to be considered together. Recent studies have unveiled the key interactions between the cell cycle and the stress response networks [2, 8, 14, 19, 55]. To build a comprehensive model of the cell cycle response of *S. cerevisiae* to osmotic stress and the α-factor, we have considered the corresponding stress response signalling networks and their coupling with the cell cycle. In the following section we will give an overview of these two networks and their interactions with the cell cycle.

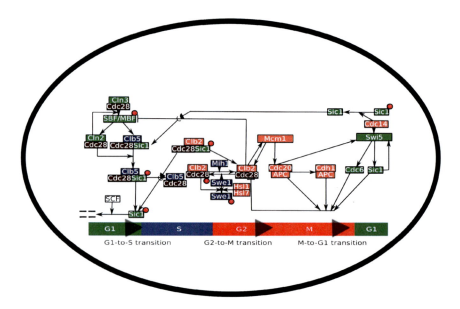

Fig. 2.6 Simplified schematic of budding yeast's cell cycle control network. As soon as cyclins are synthesised, they bind to Cdc28. As this tethering is fast, the cyclins alone have not been shown in the figure. Cell cycle control components are coloured based on the phase of activity, namely, green for G1 phase, blue for S phase and pink for G2/M phase. For details of the interactions see text

2.3 Osmotic Stress Response

Osmoregulation is a homeostatic process which controls: internal turgor pressure, the water content, and the volume of the cell. Various receptors sense osmotic stress and cause the activation of the High-Osmolarity Glycerol (HOG) MAPK signalling [9]. The HOG MAP kinase pathway, like any other MAP kinase pathway, consists of a cascade of three kinases: a MAP kinase (MAPK), a MAP kinase kinase (MAPKK), and a MAP kinase kinase kinase (MAPKKK) [21, 26].

For *S. cerevisiae* Sln1, Msb2, Hkr1 and Sho1 are osmosensors and upstream of two branches (see Fig. 2.7), which monitor changes in the turgor pressure and independently regulate three MAPKKKs (Ste11, Ssk2 and Ssk22) [26, 36, 38, 49, 54]. Consequently, MAPKKKs phosphorylate and activate the MAPKK (Pbs2) [26]. Activation of Pbs2 results in the activation of Hog1 via phosphorylation [26]. Dually phosphorylated Hog1 (Hog1PP) then accumulates in the nucleus and regulates the response of the cell to osmotic stress; (i) it mediates the production of glycerol to compensate for the turgor pressure loss [21]. (ii) Hog1PP interacts with cell cycle regulated components and block progression through the G1-to-S and the G2-to-M transitions [2, 8, 14, 19, 55]. Phosphorylated Hog1 is the main player in the interactions between the osmotic stress response and the cell cycle.

2.4 Interaction Between the Cell Cycle Network and the Osmotic Stress Pathway

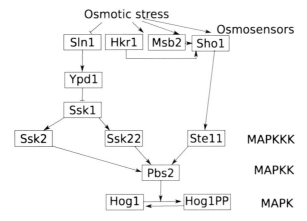

Fig. 2.7 Osmotic stress response network of *S. cerevisiae*; Sln1, Sho1 and Msb2/Hkr2 sense osmotic stress. These sensors are upstream of two branches which monitor changes of the turgor pressure and, independently, regulate three MAPKKKs (Ste11, Ssk2 and Ssk22). Subsequently, MAPKKKs phosphorylate and activate MAPKK (Pbs2). Activation of Pbs2 results in the activation of Hog1 via phosphorylation. Phosphorylated Hog1 then accumulates in the nucleus and activates gene expression of proteins involved in glycerol production. Thereby, glycerol is produced to compensate for the loss of turgor pressure

2.4 Interaction Between the Cell Cycle Network and the Osmotic Stress Pathway

In the last decade different interactions between Hog1PP and cell cycle regulated proteins leading to cell cycle arrest have been experimentally identified [2, 8, 14, 19, 55]. Crucially, the cell cycle phase during which the osmotic stress is applied determines the mechanisms of the interaction between the osmotic stress response and the cell cycle machinery.

2.4.1 Osmotic Stress Blocks the G1-to-S Transition

In untreated cells, activity of Cdc28-G1-cyclin complexes triggers the G1-to-S transition by targeting Sic1 (CKI) for degradation; thereby freeing the S phase cyclin to initiate DNA replication (see Fig. 2.2) [42]. In *S. cerevisiae*, if osmotic stress is applied during the G1 phase, activation of Hog1 results in G1 arrest by a dual mechanism (see Figs. 2.8 and 2.9) [8, 19]:

(i) Hog1PP downregulates the transcription of the G1 cyclins, leading to Sic1 accumulation [8, 19].

(ii) Hog1PP stabilises Sic1; Phosphorylation of Sic1 by Hog1PP, reduces the binding of Sic1 to the SCF complex to less than 30%. This complex formation is

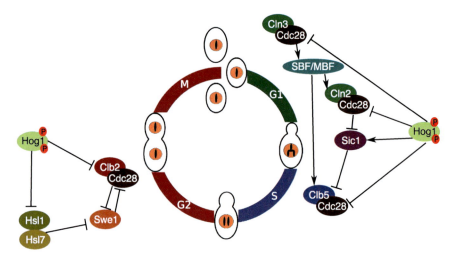

Fig. 2.8 Summary of the interactions of Hog1PP with cell cycle components. Hog1PP blocks the G1-to-S transition by phosphorylation and accumulation of Sic1; and transcriptional downregulation of G1 cyclins. Also, Hog1PP blocks the G2-to-M transition by indirect accumulation of Swe1; and transcriptional downregulation of G2/M cyclins. For details of interactions see text

required to initiate the ubiquitin-mediated proteolysis pathway responsible for Sic1 degradation [45]. Therefore, the presence of Hog1PP stabilises Sic1 [19].

Hence, the G1 arrest as a response to osmotic stress is a consequence of the direct phosphorylation of Sic1 and the downregulation of the G1 cyclins by Hog1PP.

2.4.2 Osmotic Stress Blocks the G2-to-M Transition

The progression through the S and G2/M phases are also delayed by osmotic stress due to three main mechanisms: (i) direct downregulation of *CLB5* transcription by Hog1PP [55], (ii) accumulation of Swe1 and (iii) downregulation of *CLB2* transcription (see Figs. 2.8 and 2.9).

As mentioned before, Swe1 is a part of the morphogenesis checkpoint and halts the cell cycle progression if bud formation is defective. In non-stress conditions, Swe1 is absent in G1 and accumulates during late G1 and S phases before disappearing rapidly in the G2/M phase [29, 44]. The Hsl1-Hsl7 complex and active Cdc28-Clb2 mediate the rapid degradation of Swe1 [30, 44]. Hsl1 is stable in the G1 phase, and accumulates during the S phase to peak in G2/M phase; its degradation is synchronised with the nuclear division [30]. Hsl1 changes its location between the nucleus and the bud neck, depending on the cell cycle phase. In contrast, Hsl7 shuttles between the spindle pole body and the bud neck and its total levels do not change significantly during the cell cycle [13, 30].

2.4 Interaction Between the Cell Cycle Network and the Osmotic Stress Pathway

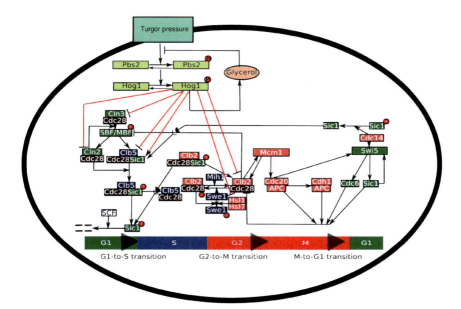

Fig. 2.9 Summary of stress signalling pathway; cell cycle control network; and interactions between them. Cells sense the osmotic stress and activate signalling pathway which results in the activity and phosphorylation of Hog1. Hog1PP regulates the loss of turgor pressure and the progression through the cell cycle. Cell cycle control components are coloured based on the cell cycle phase when they are active, namely, green for G1 phase, blue for S phase and pink for G2/M phase. The interaction between the Hog1PP and the cell cycle components are shown by red links. For details, see text

However, in the presence of osmotic stress, Hog1PP targets Hsl1 for phosphorylation [14], and hinders the Hsl1-Hsl7 complex formation. As a consequence, Swe1 is not degraded and Hsl7 is delocalised from the bud neck [14]. Hence, osmotic stress applied at the onset of the G2 phase inhibits Cdc28-Clb2 as a consequence of Swe1 stabilisation and the direct downregulation of *CLB2*. This leads to the G2 arrest.

In many experimental setups a population of cells needs to be synchronised– to study the influence of stress at a certain stage of the cell cycle. One method of synchronisation is "block-and-release". In this method a population of cells is arrested at a specific cell cycle stage and released from the block; subsequently, cells are sampled at different time points after release [20]. This type of synchronisation alters the dynamics of the cell cycle regulation. For example, application of the α-factor to yeast cells arrests them in the G1 phase before release. To study the influence of the α-factor arrest on the cell cycle dynamics of untreated and treated with osmotic stress culture, we will next describe (i) the mating-pheromone signalling pathway, and (ii) the interactions between this signalling pathway and the cell cycle.

2.5 Mating-Pheromone Response

S. cerevisiae cells can exist as either diploid (cells have two homologous copies of each chromosome) or haploid (cells have one copy of each chromosome). The mating of yeast only occurs between two types of haploid yeast; $MATa$ and $MAT\alpha$ cells. Opposite mating type yeasts ($MATa$ and $MAT\alpha$) mate by secreting pheromone (e.g., α-factor); presence of pheromone triggers a cascade of events to prepare the cell for mating, such as: (i) changes in the expression of about 200 genes [5]; (ii) and arrest in the G1 phase of the cell cycle [5].

S. cerevisiae senses the presence of extracellular pheromone by a signal transduction network called mating pathway [5]. The mating pathway is a MAPK signalling network and one of the best understood signalling pathways in eukaryotes [5]. Upstream of this pathway there are cell-surface receptors (Ste2, Ste3) to which the mating pheromone binds (see Fig. 2.10) [10, 23]. For the mating pathway of *S. cerevisiae*, the MAPKKK is Ste11; the MAPKK is Ste7; and there are two MAPKs, Kss1 and Fus3 [5].

External pheromone is sensed by receptors. This leads to the regulation of Ste11. Consequently, Ste11 phosphorylates and activates Ste7 (see Fig. 2.10). Activation of Ste7 results in the activation of Kss1 and Fus3 via phosphorylation. Activation of Fus3 regulates pheromone-induced gene expression and phosphorylates Far1, which

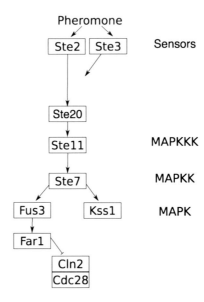

Fig. 2.10 Pheromone stress response network of *S. cerevisiae*. An external pheromone is sensed by Ste2 and Ste3. They are upstream of the MAPKKK (Ste11). Consequently, MAPKKKs phosphorylate and activate MAPKK (Ste7). Activation of Ste7 results in the activation of Kss1 and Fus3 via phosphorylation. Phosphorylated Fus3 then phosphorylates Far1, which induces cell cycle arrest

regulates the mating process in various ways. For example, Far1 mediates the cell cycle arrest [11, 51].

2.5.1 Pheromone Arrests the Cell Before START

As explained above, Cdc28 is the main regulator of the cell cycle of *S. cerevisiae*. Far1 is one of the inhibitors (CKIs) of Cdc28 and can inactivate Cdc28-Cln2 (see Fig. 2.10); Therefore, it blocks the cell cycle progression before start [11, 37].

Since the pheromone arrests cells in the G1 phase, it is frequently used to synchronise a population of cells at the START. This method of synchronisation is called α-factor synchronisation and is one of the "block-and-release" synchronisation methods [20].

Despite all these known interactions between the stress signalling pathway and the cell cycle network (see Figs. 2.9 and 2.10), many more aspects of the cell cycle response to stresses are still unknown. For example: (i) the response of cells in the S and M phases at the onset of stress; (ii) the influence of various levels of stress on the cell cycle; (iii) the main interactions in response to osmotic stress; (iv) the response of different mutated cells to osmotic stress; (v) the dynamics of the cell cycle regulation in the presence of stress; and (vi) the impact of "block-and-release" synchronisation on cell cycle dynamics, are unknown.

As we argued in this Chapter, the complexity of cell cycle machinery and stress response network (see Fig. 2.9) defies intuition and requires a systems biology approach. Hence, to study the regulation of all cell cycle events, which are controlled by an interconnected network, in the presence of extracellular signals, we present two comprehensive and complementary mathematical models [39, 40]. These models include all known interactions between the cell cycle and these two extracellular signals and answer these questions, for the first time.

References

1. B. Alberts, D. Bray, K. Hopkin, A. Johnson, J. Lewis, K. Roberts, M. Raff and P. Walter. *Essential cell biology* 2nd edn. (Taylor and Francis 2003)
2. M.R. Alexander, M. Tyers, M. Perret, B.M. Craig, K.S. Fang, M.C. Gustin, Regulation of cell cycle progression by Swe1p and Hog1p following hypertonic stress. Mol. Biol. Cell **12**(1), 53–62 (2001)
3. A. Amon, M. Tyers, B. Futcher, K. Nasmyth, Mechanisms that help the yeast cell cycle clock tick: G2 cyclins transcriptionally activate G2 cyclins and repress G1 cyclins. Cell **74**(6), 993–1007 (1993)
4. S. Asano, J.-E. Park, K. Sakchaisri, L.-R. Yu, S. Song, P. Supavilai, T.D. Veenstra, K.S. Lee, Concerted mechanism of Swe1/Wee1 regulation by multiple kinases in budding yeast. EMBO J. **24**(12), 2194–2204 (2005)
5. L. Bardwell, A walk-through of the yeast mating pheromone response pathway. Pept. **26**(2), 339–350 (2005)

6. R.D. Basco, M.D. Segal, S.I. Reed, Negative regulation of G1 and G2 by S-phase cyclins of Saccharomyces cerevisiae. Mol. Cell. Biol. **15**(9), 5030–5042 (1995)
7. M. Bäumer, G.H. Braus, S. Irniger, Two different modes of cyclin Clb2 proteolysis during mitosis in Saccharomyces cerevisiae. FEBS Lett. **468**(2–3), 142–148 (2000)
8. G. Bellí, E. Garí, M. Aldea, E. Herrero, Osmotic stress causes a G1 cell cycle delay and downregulation of Cln3/Cdc28 activity in Saccharomyces cerevisiae. Mol. Microbiol. **39**(4), 1022–1035 (2001)
9. J.L. Brewster, T. De Valoir, N.D. Dwyer, E. Winter, M.C. Gustin, An osmosensing signal transduction pathway in yeast. Sci. **259**(5102), 1760–1763 (1993)
10. A.C. Burkholder, L.H. Hartwell, The yeast alpha-factor receptor: Structural properties deduced from the sequence of the STE2 gene. Nucleic Acids Res. **13**(23), 8463–8475 (1985)
11. F. Chang, I. Herskowitz, Identification of a gene necessary for cell cycle arrest by a negative growth factor of yeast: FAR1 is an inhibitor of a G1 cyclin, CLN2. Cell **63**(5), 999–1011 (1990)
12. G. Charvin, C. Oikonomou, E.D. Siggia and F.R. Cross. Origin of irreversibility of cell cycle start in budding yeast. PLoS Biol. **8**(1), e1000284 (2010)
13. V.J. Cid, M.J. Shulewitz, K.L. McDonald, J. Thorner, Dynamic localization of the Swe1 regulator Hsl7 during the Saccharomyces cerevisiae cell cycle. Mol. Biol. Cell **12**(6), 1645–1649 (2001)
14. J. Clotet, X. Escoté, M.A. Adrover, G. Yaakov, E. Garí, M. Aldea, E. de Nadal, F. Posas, Phosphorylation of Hsl1 by Hog1 leads to a G2 arrest essential for cell survival at high osmolarity. EMBO J. **25**(11), 2338–2346 (2006)
15. C. Dahmann, J.F. Diffley, K.A. Nasmyth, S-phase-promoting cyclin-dependent kinases prevent re-replication by inhibiting the transition of replication origins to a pre-replicative state. Curr. Biol. **5**(11), 1257–1269 (1995)
16. J.F.X. Diffley, Once and only once upon a time: Specifying and regulating origins of DNA replication in eukaryotic cells. Genes. Devel. **10**(22), 2819–2830 (1996)
17. L. Dirick, T. Böhm, K. Nasmyth, Roles and regulation of Cln-Cdc28 kinases at the start of the cell cycle of Saccharomyces cerevisiae. EMBO J. **14**(19), 4803–4813 (1995)
18. S.J. Elledge, Cell cycle checkpoints: Preventing an identity crisis. Sci. **274**(5293), 1664–1672 (1996)
19. X. Escoté, M. Zapater, J. Clotet, F. Posas, Hog1 mediates cell-cycle arrest in G1 phase by the dual targeting of Sic1. Nat. Cell Biol. **6**(10), 997–1002 (2004)
20. B. Futcher, Cell cycle synchronization. Methods Cell Sci. **21**(2–3), 79–86 (1999)
21. M.C. Gustin, J. Albertyn, M. Alexander, K. Davenport, Map kinase pathways in the yeast Saccharomyces cerevisiae. Microbiol. Mol. Biol. Rev. **62**(4), 1264–1300 (1998)
22. J.A. Hadwiger, C. Wittenberg, H.E. Richardson, M. De Barros Lopes, S.I. Reed, A family of cyclin homologs that control the G1 phase in yeast. PNAS **86**(16), 6255–6259 (1989)
23. D.C. Hagen, G. McCaffrey, G.F. Sprague Jr, Evidence the yeast STE3 gene encodes a receptor for the peptide pheromone a factor: Gene sequence and implications for the structure of the presumed receptor. PNAS **83**(5), 1418–1422 (1986)
24. L.H. Hartwell and M.W. Unger. Unequal division in Saccharomyces cerevisiae and its implications for the control of cell division. J. Cell Biol. **75**(2) 422–435(1977)
25. L.H. Hartwell, T.A. Weinert, Checkpoints: controls that ensure the order of cell cycle events. Sci. **246**(4930), 629–634 (1989)
26. S. Hohmann, Osmotic stress signaling and osmoadaptation in yeasts. Microbiol. Mol. Biol. Rev. **66**(2), 300–372 (2002)
27. P. Jorgensen, I. Rupees, J.R. Sharom, L. Schneper, J.R. Broach, M. Tyers, A dynamic transcriptional network communicates growth potential to ribosome synthesis and critical cell size. Genes. Dev. **18**(20), 2491–2505 (2004)
28. D.J. Lew, S.I. Reed, Morphogenesis in the yeast cell cycle: regulation by Cdc28 and cyclins. J. Cell Biol. **120**(6), 1305–1320 (1993)
29. D.J. Lew, Cell-cycle checkpoints that ensure coordination between nuclear and cytoplasmic events in Saccharomyces cerevisiae. Curr. Opin. Genet. Dev. **10**(1), 47–53 (2000)

References

30. J.N. McMillan, M.S. Longtine, R. Sia, C.L. Theesfeld, E.S. Bardes, J.R. Pringle, D.J. Lew, The morphogenesis checkpoint in Saccharomyces cerevisiae: cell cycle control of Swe1p degradation by Hsl1p and Hsl7p. Mol. Cell. Biol. **19**(10), 6929–6939 (1999)
31. M.D. Mendenhall, A.E. Hodge, Regulation of Cdc28 cyclin-dependent protein kinase activity during the cell cycle of the yeast Saccharomyces cerevisiae. Microbiol. Mol. Biol. Rev. **62**(4), 1191–1243 (1998)
32. A. Murray and T. Hunt. *the cell cycle*: An introduction (Oxford University Press, New York, 1993)
33. P. Nash, X. Tang, S. Orlicky, Q. Chen, F.B. Gertler, M.D. Mendenhall, F. Sicheri, T. Pawson, M. Tyers, Multisite phosphorylation of a CDK inhibitor sets a threshold for the onset of DNA replication. Nat. **414**(6863), 514–521 (2001)
34. E.A. Nigg, Mitotic kinases as regulators of cell division and its checkpoints. Nat. Rev. Mol. Cell Biol. **2**(1), 21–32 (2001)
35. P. Nurse, Regulation of the eukaryotic cell cycle. Eur. J. Cancer Part A **33**(7), 1002–1004 (1997)
36. S.M. O'Rourke and I. Herskowitz. A third osmosensing branch in Saccharomyces cerevisiae requires the Msb2 potein and functions in parallel with the Sho1 branch. Mol. Cell. Bio. **22**(13), 4739–4749 (2002)
37. M. Peter, I. Herskowitz, Direct inhibition of the yeast cyclin-dependent kinase Cdc28-Cln by Far1. Sci. **265**(5176), 1228–1231 (1994)
38. F. Posas, H. Saito, Osmotic Activation of the HOG MAPK Pathway via Ste11p MAPKKK: Scaffold Role of Pbs2p MAPKK. Sci. **276**(5319), 1702–1705 (1997)
39. E. Radmaneshfar, M. Thiel, Recovery from stress: a cell cycle perspective. J. Comp. Int. Sci. **3**(1–2), 33–44 (2012)
40. E. Radmaneshfar, D. Kaloriti, M.C. Gustin, N.A.R Gow, A.J.P Brown, C. Grebogi, M.C. Romano and M. Thiel. From START to FINISH: the influence of osmotic stress on the cell cycle. PLoS ONE **8**(7), e68067 (2013)
41. H.E. Richardson, C. Wittenberg, F. Cross, S.I. Reed, An essential G1 function for cyclin-like proteins in yeast. Cell **59**(6), 1127–1133 (1989)
42. E. Schwob, T. Böhm, M.D. Mendenhall, K. Nasmyth, The B-type cyclin kinase inhibitor p40SIC1 controls the G1 to S transition in S. cerevisiae. Cell **79**(2), 233–244 (1994)
43. R. Sia, H. Herald, D.J. Lew, Cdc28 tyrosine phosphorylation and the morphogenesis checkpoint in budding yeast. Mol. Biol. Cell **7**(11), 1657–1666 (1996)
44. R. Sia, E.S. Bardes, D.J. Lew, Control of Swe1p degradation by the morphogenesis checkpoint. EMBO J. **17**(22), 6678–6688 (1998)
45. D. Skowyra, K.L. Craig, M. Tyers, S.J. Elledge, J.W. Harper, F-box proteins are receptors that recruit phosphorylated substrates to the SCF ubiquitin-ligase complex. Cell **91**(2), 209–219 (1997)
46. P.T. Spellman, G. Sherlock, M.Q. Zhang, V.R. Iyer, K. Anders, M.B. Eisen, P.O. Brown, D. Botstein, B. Futcher, Comprehensive identification of cell cycle-regulated genes of the yeast Saccharomyces cerevisiae by microarray hybridization. Mol. Biol. Cell **9**(12), 3273–3297 (1998)
47. F. Stegmeier, A. Amon, Closing mitosis: The functions of the Cdc14 phosphatase and its regulation. Annu. Rev. Genet. **38**, 203–232 (2004)
48. U. Surana, H. Robitsch, C. Price, T. Schuster, I. Fitch, B. Futcher, K. Nasmyth, The role of CDC28 and cyclins during mitosis in the budding yeast S. cerevisiae. Cell **65**(1), 145–161 (1991)
49. K. Tatebayashi, K. Tanaka, H.-Y. Yang, K. Yamamoto, Y. Matsushita, T. Tomida, M. Imai, H. Saito, Transmembrane mucins Hkr1 and Msb2 are putative osmosensorsy in the SHO1 branch of yeast HOG pathway. EMBO J. **26**(15), 3521–3533 (2007)
50. J.H. Toyn, A.L. Johnson, J.D. Donovan, W.M. Toone and L.H. Johnstone. The Swi5 transcription factor of Saccharomyces cerevisiae has a role in exit from mitosis through induction of the cdk-inhibitor Sic1 in telophase. Genet. **145**(1) 85–96 (1997)
51. M. Tyers, B. Futcher, Far1 and Fus3 link the mating pheromone signal transduction pathway to three G1-phase Cdc28 kinase complexes. Mol. Cell. Biol. **13**(9), 5659–5669 (1993)

52. M. Tyers, G. Tokiwa, B. Futcher, Comparison of the Saccharomyces cerevisiae G1 cyclins: Cln3 may be an upstream activator of Cln1, Cln2 and other cyclins. EMBO J. **12**(5), 1955–1968 (1993)
53. R. Visintin, S. Prinz, A. Amon, CDC20 and CDH1: a family of substrate-specific activators of APC- dependent proteolysis. Sci. **278**(5337), 460–463 (1997)
54. P.J. Westfall, D.R. Ballon, J. Thorner, When the stress of your environment makes you go HOG wild. Sci. **306**(5701), 1511–1512 (2004)
55. G. Yaakov, A. Duch, M. Garcí-Rubio, J. Clotet, J. Jimenez, A. Aguilera, F. Posas, The stress-activated protein kinase Hog1 mediates S phase delay in response to osmostress. Mol. Biol. Cell **20**(15), 3572–3582 (2009)

Chapter 3
ODE Model of the Cell Cycle Response to Osmotic Stress

3.1 Introduction

Cells live in changing environment; they sense and activate molecular signalling networks to respond to external signals. Cell cycle accordingly coordinates its tightly controlled biochemical events with the activity of signalling networks. To understand such a complex system we have developed a mathematical model which integrates the osmotic stress response network with the cell cycle machinery of budding yeast, *S. cerevisiae*.

The first published model focused on the interaction between the osmotic stress response and only the G1 phase of the cell cycle of *S. cerevisiae* [1]. Yet, since the cell cycle phases are intertwined by control mechanisms, the effect of osmotic stress on the cell cycle cannot be predicted from the consideration of one single phase alone. In this chapter we introduce a comprehensive mathematical model that, for the first time, describes the effect of osmotic stress throughout the whole cell cycle [45]. Our mathematical model integrates a large number of data sets and condenses the biological knowledge gained from those experiments (reviewed in Chap. 2) into mathematical equations and elucidates how this elaborate system might work in the presence of osmotic stress. The model generates a series of novel predictions. It predicts the influence of osmotic stress on the progression through the S and M phases of the cell cycle; which is experimentally unknown. It also provides a tool for further investigation of the molecular processes and the cell behaviour under various environmental conditions and experimental setups. The model analyses the influence of different doses of osmotic stress on the progression through the entire cell cycle. Some of the predictions suggests the molecular mechanisms of experimental observations reported recently in the literature [47]; thereby validating the predictive power of the model.

The molecular basis of cell cycle control is highly conserved among different eukaryotes. Hence, the predictions by our integrated mathematical model of the cell cycle and stress response in *S. cerevisiae* are relevant for other eukaryotes. The recent study validates this postulation and also confirms the predictive power of the model.

E. Radmaneshfar, *Mathematical Modelling of the Cell Cycle Stress Response*,
Springer Theses, DOI: 10.1007/978-3-319-00744-1_3,
© Springer International Publishing Switzerland 2014

28   3   ODE Model of the Cell Cycle Response to Osmotic Stress

Next, we first explain our approach in modelling the response of the cell cycle to osmotic stress; second we analyse the model and mention the predictions of model; third we discuss the biological evidences that support the predictions of the model and finally we will elucidate the relevance of the prediction of this model for other eukaryotes.

3.2 Modelling Procedure and Assumptions

3.2.1 Steps to Construct the Model

Our modelling approach consists of four major steps:

(i) Several molecular mechanisms which control the cell cycle of *S. cerevisiae*; and affects its regulation in the presence of osmotic stress have been reported in the literature (reviewed in Chap. 2) [2, 3, 7, 8, 10, 17, 21, 24, 33, 34, 39, 41, 48, 50–52, 57–59, 61, 62]. Based on this information and other experimental evidence [9, 41, 55, 63], we construct the wiring diagram depicted in Fig. 3.1 [45]. The components presented in this figure are involved in: the cell cycle regulation, and the response of the cell cycle to osmotic stress. The first group are coloured based on their phase of activity, namely, green for G1 phase, blue for S phase and pink for G2/M phase. The second group are coloured in orange if they have been explicitly reported to be involved in the response of the cell cycle to osmotic stress [2, 8, 17, 24]; and in yellow if there are experimental evidences that indicate their participation in the response of the cell cycle to osmotic stress [9, 41, 55].

(ii) We formulate the mechanisms which control the interaction between the cell cycle and the response to osmotic stress in the form of mathematical equations. In general, the time profile of the concentration $[C]$ of each component depends on the sum of its production/activation rates $v_{p/a}$, and the sum of its degradation/inhibition rates $v_{d/i}$:

$$\frac{d[C]}{dt} = \sum v_{p/a} - \sum v_{d/i}, \qquad (3.1)$$

where $\sum v_{p/a}$ and $\sum v_{d/i}$ depend on the kinetics of the corresponding interactions.

We use mass-action kinetics, Michaelis-Menten kinetics, and Hill functions to describe the production/activation and degradation/inhibition interactions of component C with other components. For components that are either active or inactive during the cell cycle, like the transcription factors SBF, MBF and Mcm1, we use a switch like function called "Goldbeter-Koshland function" [26]. Note that there is no unique way to translate the wiring diagram into equations, and the final model depends on the level of required details.

The model has two main modules: the cell cycle module and the osmotic stress module. The cell cycle module is based on the successful model developed by Chen et al. [13]. The extension of this model consists in the introduction of the links

3.2 Modelling Procedure and Assumptions

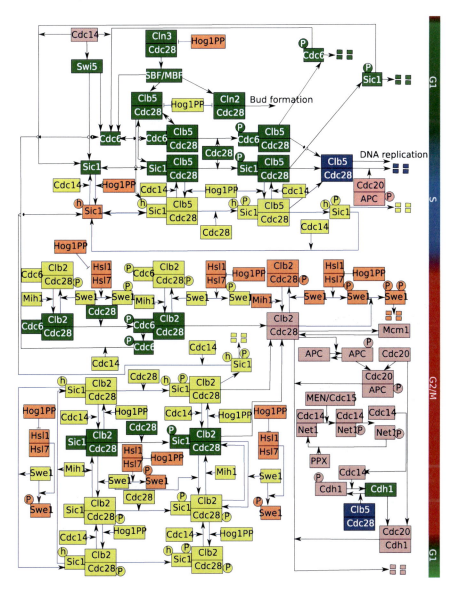

Fig. 3.1 Wiring diagram of the interactions of Hog1PP with the cell cycle components. This network summarises the biological interactions and modelling assumptions. Active Hog1PP halts cell cycle progression depending on the phase of the cell cycle at the onset of osmotic stress. The components marked in *yellow and orange* in this figure represent elements that couple the cell cycle with the osmotic stress response network. Components involved in known biological interactions are marked in *orange* and the ones involved in assumed interactions based on reported experimental evidence are marked in *yellow* (see Sect. 3.2.2 for details of modelling assumptions). Cell cycle control components are coloured based on the phase of activity, namely, *green* for G1 phase, *blue* for S phase and pink for G2/M phase (see Chap. 2 for details of cell cycle control interactions)

between Hog1PP and different components of the cell cycle machinery. The osmotic stress module, on the other hand, is based on the model developed by Zi et al. [65] for the activation of Hog1PP.

First, new cell cycle elements are added to the cell cycle module model, namely Swe1, Hsl1, Hsl7, Mih1 and their complexes. These are the cell cycle regulated components which are affected by Hog1PP but are absent in the model of Chen et al. [13]. Second, the influence of Hog1PP activity on the regulation of the targeted cell cycle components are modelled. To our knowledge no feedback from the cell cycle machinery to the osmotic stress response has been reported. Hence, we keep the interaction unidirectional. Details of the modelling assumptions are explained in Sect. 3.2. For the complete list of equations, see the Appendix A.

(iii) After translating the proposed wiring diagram (Fig. 3.1) into mathematical equations (see Appendix A), the model is parameterised. A large number of the parameters are taken from the literature [12, 13, 15, 18, 21, 23–25, 56, 64, 65]. The other parameters are chosen to reproduce the behaviour of various mutants, as well as the cell cycle arrest duration upon application of osmotic stress in different experimental setups (see Sect. 3.4). The sensitivity analysis shows that the predictions of the model are robust against changes of the parameters. For details of estimation of parameters and also sensitivity analysis, see Sect. 3.4.

(iv) Finally the set of equations were numerically solved by a 4th order Runge-Kutta algorithm in MATLAB (MathWorks).

The resulting model encompasses a vast amount of known biological experimental knowledge; it adds a substantial amount of information by making biological implicit assumptions mathematically explicit.

3.2.2 Hypothesised Mechanisms in the Model

First, we assume that Hog1PP can phosphorylate Sic1 when the latter is in any of its forms, that means, (i) when Sic1 is in a complex or (ii) it has been already phosphorylated by Cdc28 or (iii) when it is unbound. Since, Sic1 has 9 phosphorylation sites [41] and the site at which Hog1PP phosphorylates Sic1 is different from the site at which Cdc28 phosphorylates Sic1 [24], this assumption is plausible. Cdc28-phosphorylated Sic1 is denoted by Sic1P and Hog1PP-phosphorylated Sic1 is denoted by Sic1h in the equations of the model and Figs. 3.1, 3.2a, b. Nevertheless, given the experimental evidence that Sic1h has a reduced binding affinity to the SCF complex [24], Sic1h is assumed to be more stable than Sic1P in the model, both in free and in complex form.

Second, it is also reasonable to assume that Sic1 can be first phosphorylated by Hog1PP and then by Cdc28, both when Sic1 is not associated with Cdc28 and when it is bound in a complex.

Third, we assume that Hog1PP-phosphorylated Sic1 (Sic1h) can also bind to the Cdc28-B-type cyclin complexes.

3.2 Modelling Procedure and Assumptions

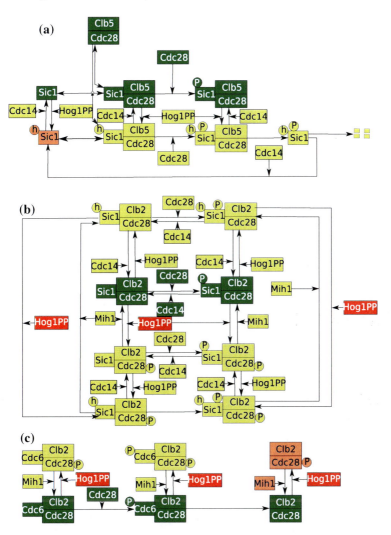

Fig. 3.2 Schematic diagram of modelling assumptions for interaction of Hog1PP with cell cycle regulated components. *Green blocks* show the regulation of the component when Hog1PP is absent. *Orange and red blocks* represent the reported components that are involved in the response to osmotic stress. The Hog1PP, which is shown by the *red block*, is the box of interactions of Hog1PP with the Hsl1-Hsl7 complex that ends up in the accumulation of Swe1. These interactions are considered in the model, but it is shown in a box here for clarity purpose of the figure. The *yellow blocks* show the assumed mechanisms in response to osmotic stress. The Hog1PP-phosphorylated Sic1 is denoted by Sic1h and Cdc28-phosphorylated Sic1 is denoted by Sic1P. See text for biological evidence of the assumptions. (**a**) The assumed response mechanisms of the G1 phase of the cell cycle to activation of Hog1PP. (**b**) and (**c**) The assumed response mechanisms of the cell cycle to activation of Hog1PP before entry into mitosis. Note that the downregulation of cyclins by Hog1PP has not been shown in this figure

Forth, we assume that Hog1PP-phosphorylated Sic1 (Sic1h) can be dephosphorylated by the phosphatase Cdc14; it is known that Cdc14 dephosphorylates Cdc28-phosphorylated Sic1 (Sic1P) and almost all substrates involved in the G1-to-S transition [55].

Finally, it is known that Swe1 phosphorylates and inhibits Cdc28-Clb2 [9]. Likewise, we assume that Swe1 can phosphorylate any complex containing Cdc28-Clb2 [15].

All these interactions are summarised in Fig. 3.2a–c.

3.2.3 Mathematical Definition of the Cell Cycle Phases

The precise experimental determination of the limits between different cell cycle phases is not straightforward; often cell cycle phases are determined in single cells by monitoring bud formation via microscopy, or DNA content in fluorescence-activated cell sorting (FACS) experiments in the case of a synchronised cell population. The cell cycle, however, is a continuous rather than a discrete progression of biochemical events. When referring to the model predictions though, it is useful to have a precise definition of the borders between the phases. Therefore, here we introduce a mathematical definition of the limits between the cell cycle phases, which we use throughout the paper: the G1 phase starts right after cell division, and finishes when the level of Cdc28-Clb5 crosses the level of Sic1 [4, 41]. This indicates initiation of DNA replication and therefore the start of the S phase. The S phase finishes when the level of Cdc28-Clb2 becomes greater than the level of Swe1 [34], also defining the start of the G2/M phase (see Fig. 3.3). The end of the G2/M phase is defined as the point at which Cdc28-Clb2 becomes less than Sic1, which indicates cell division [13]. A further key biochemical event is defined by Mcm1 reaching its maximum level, which marks the beginning of the FINISH process [12]. Note that these definitions of the limits between different cell cycle phases use the cell cycle of a non-stressed cell as a reference. Under osmotic stress the biochemical events dictating the transitions between the phases are distorted and therefore the chosen definitions serve just as reference points (see Fig. 3.3).

3.3 Model Description

3.3.1 The Morphogenesis Checkpoint

The morphogenesis checkpoint refers to the control mechanisms that hinder the progression through the G2 phase under unfavourable conditions, such as defective bud formation or environmental stress. This mechanism is achieved through the control of Cdc28-Clb2 activity [50, 51]. This checkpoint is inactive under optimal

3.3 Model Description

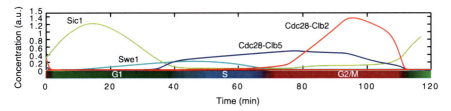

Fig. 3.3 Mathematical definition of the cell cycle phases. The G1 phase starts immediately after cell division, and finishes when the level of Cdc28-Clb5 exceeds the level of Sic1, which indicates initiation of DNA replication and the start of the S phase. The S phase finishes when the level of Swe1 is less than the level of Cdc28-Clb2. Note that we do not distinguish between G2 and M phase, as G2 phase is very short for *S. cerevisiae*

growth conditions, but malformation of bud or osmotic stress activate this checkpoint, and block the cell cycle progression. The morphogenesis checkpoint was absent in the cell cycle model by Chen et al. [13].

Activity of Swe1 during late G1 and S phase inhibits Cdc28-Clb2 activity. In turn, the activity of Swe1 is blocked by the activity of Hsl1-Hsl7 and also by Cdc28-Clb2 complexes. We have developed a model for the morphogenesis checkpoint, which includes the dynamics of Hsl1-Hsl7 regulations explicitly (for details of regulations of Hsl1-Hsl7 complex see Sect. 3.3.3), considering the recently unveiled molecular mechanisms of the morphogenesis checkpoint [10, 34, 52, 58]. This is in contrast to the model of Ciliberto et al., where the dynamics of Hsl1-Hsl7 was not considered and instead reduced to a fixed parameter [15]. We assume that Swe1 can exist in four different forms, namely, unphosphorylated (denoted by Swe1), phosphorylated by Hsl1-Hsl7 (Swe1M), phosphorylated by Cdc28-Clb2 (denoted by Swe1P), and doubly phosphorylated (denoted by Swe1MP) in accordance with biological studies [9, 33, 34, 36, 50, 51, 58] (see Fig. 3.4). Note that the last form is expected to be highly unstable [37]. The morphogenesis checkpoint model includes the components Swe1, Mih1, Hsl1, Hsl7, Cdc28-Clb2 and, additionally, it incorporates the interactions that link the morphogenesis checkpoint to the osmotic stress response module. Note that the mechanisms responsible for the morphogenesis checkpoint are highly complex and, in our model we do not include components such as Gin4, Elm1, Cla4, which have been implicated in the morphogenesis checkpoint, but do not interact with Hog1PP. As such, we keep the model as simple as possible, while preserving the main biological mechanisms. We then integrate this model into the cell cycle module by considering the interaction of the morphogenesis checkpoint elements with the rest of the cell cycle components.

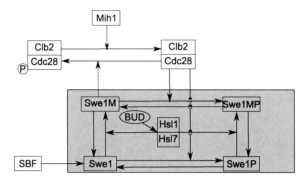

Fig. 3.4 Simplified schematic diagram of morphogenesis checkpoint: Swe1 kinase phosphorylates Cdc28-Clb2 and halts the transition to the G2/M phase. *SWE1* transcription is by the transcription factor SBF. It needs to be phosphorylated in order to become degraded. This phosphorylation occurs via Cdc28-Clb2 and the Hsl1-Hsl7 complex. If the modification is caused by Cdc28-Clb2, the product is denoted by Swe1P in the diagram, whereas the product that is mediated by the Hsl1-Hsl7 complex is denoted by Swe1M. The substrate Swe1MP is highly unstable. We assume that the total concentration of Swe1 inhibits the activity of Cdc28-Clb2; see text for further details. The formation of the complex Hsl1-Hsl7 coincides with the formation of the bud. This timing is shown by the variable denoted by BUD in the figure. BUD denotes a mathematical function which describes the observation of bud formation and depends on Cln2, Cln3 and Clb5. The degradation of Hsl1-Hsl7 is APC (Anaphase-Promoting Complex) – dependent, which is not shown in this figure, but has been considered in the model

3.3.2 Regulation of Swe1

Swe1 is not a crucial component for cells growing in optimal conditions: the cell cycle of *swe1*Δ cells is comparable to wild-type cells in ideal growth environment [9]. However, Swe1 becomes pivotal in cells exposed to osmotic stress and *swe1*Δ cells do not arrest in G2 phase upon the activation of the HOG MAPK [17].

We assume that Swe1 can be present in four different forms during the cell cycle in accordance with biological studies (see Fig. 3.4) [34, 36, 50, 51, 58]. Activity of Swe1 is regulated by several mechanisms:

(i) Transcription factor SBF mediates synthesis of Swe1 [50]. In wild type cells, accumulation of Swe1 begins in late G1 and peaks in S phase or early G2 phase. The production of Swe1 is described by:

$$v_{pSwe1} = k_{sswe}[SBF] + k_{ssweC}, \tag{3.2}$$

where k_{sswe} represents the synthesis rate of Swe1 by SBF and k_{ssweC} denotes the basal production rate of Swe1.

(ii) Swe1 is modified for rapid degradation during G2 and M phase by the Hsl1-Hsl7 complex and Cdc28-Clb2 complex [33, 34, 38, 51]. When the Hsl1-Hsl7 complex is present, the kinase Swe1, which has been accumulated in the nucleus by then, is moved to the bud neck [34, 38, 58]. The product of post-translational modification

3.3 Model Description

of Swe1 by Hsl1-Hsl7 is denoted by Swe1M (see Fig. 3.4 and Eq. 3.6). We model the inhibition of Swe1 by the Hsl1-Hsl7 complex with a kinetic law similar to the Hill function:

$$v_{pSwe1M} = \frac{k_{hsl1}[Hsl1Hsl7][Swe1]}{J_{iwee} + [Swe1]}, \tag{3.3}$$

where k_{hsl1} represents the rate of formation of Swe1M and J_{iwee} is the inverse of the inflection point. The formation of the Hsl1-Hsl7 complex is explained in Sect. 3.3.3.

(iii) Cdc28-Clb2 phosphorylates Swe1 [38]. We denote this intermediate product by Swe1P (Eq. 3.7) which is targeted for degradation [51]. We model the phosphorylation of Swe1 by Cdc28-Clb2 also by a Hill function since they are antagonistic:

$$v_{pSwe1P} = \frac{V_{iwee}[Clb2][Swe1]}{J_{iwee} + [Swe1]}. \tag{3.4}$$

Therefore, the dynamics of Swe1 is modelled by the following equation:

$$\frac{d[Swe1]}{dt} = v_{pSwe1} - v_{pSwe1M} + v_{rSwe1M} \tag{3.5}$$

$$- v_{pSwe1P} + v_{rSwe1P} - v_{dSwe1}$$

$$= k_{sswe}[MBF] + k_{ssweC}$$

$$- \frac{k_{hsl1}[Hsl1Hsl7][Swe1]}{J_{iwee} + [Swe1]} + k_{hsl1r}[Swe1M] - \frac{V_{iwee}[Clb2][Swe1]}{J_{iwee} + [Swe1]}$$

$$+ \frac{V_{awee}[Swe1P]}{J_{awee} + [Swe1P]} - kk_{dswe}[Swe1],$$

where v_{rSwe1M} describes the backward reaction from Swe1M to Swe1, v_{rSwe1P} corresponds to the reverse reaction from Swe1P to Swe1 and v_{dSwe1} represents the natural degradation of Swe1.

Using similar arguments, we derive the following mathematical equations describing the dynamics of Swe1M, Swe1P and Swe1MP (see Fig. 3.4):

$$\frac{d[Swe1M]}{dt} = \frac{k_{hsl1}[Hsl1Hsl7][Swe1]}{J_{iwee} + [Swe1]} - k_{hsl1r}[Swe1M] \tag{3.6}$$

$$- \frac{V_{iwee}[Clb2][Swe1M]}{J_{iwee} + [Swe1M]}$$

$$+ \frac{V_{awee}[Swe1MP]}{J_{awee} + [Swe1MP]} - kk_{dswe}[Swe1M].$$

$$\frac{d[Swe1P]}{dt} = -\frac{k_{hsl1}[Hsl1Hsl7][Swe1P]}{J_{iwee} + [Swe1P]} + k_{hsl1r}[Swe1MP] \tag{3.7}$$

$$- \frac{V_{awee}[Swe1P]}{J_{awee} + [Swe1P]}$$
$$- kk_{dswe}[Swe1P] + \frac{V_{iwee}[Clb2][Swe1]}{J_{iwee} + [Swe1]}.$$

$$\frac{d[Swe1MP]}{dt} = \frac{k_{hsl1}[Hsl1Hsl7][Swe1P]}{J_{iwee} + [Swe1P]} - k_{hsl1r}[Swe1MP] \qquad (3.8)$$
$$- \frac{V_{awee}[Swe1MP]}{J_{awee} + [Swe1MP]}$$
$$+ \frac{V_{iwee}[Clb2][Swe1M]}{J_{iwee} + [Swe1M]} - kk_{dswe}[Swe1MP].$$

We assume that the phosphorylation of Swe1 by Cdc28-Clb2 decreases the activity of the kinase Swe1 by 90% [15]. Both Swe1M (modification of Swe1 by Hsl1-Hsl7) and Swe1P (modification of Swe1 by Cdc28-Clb2) are stable, but the product Swe1MP (modification of Swe1 by Hsl1-Hsl7 and Cdc28-Clb2) is highly unstable [15]. In accordance with the reported experiments we assume that for degradation of Swe1, both Hsl1-Hsl7 and Cdc28-Clb2 are required [14, 33, 38, 51].

3.3.3 Regulation of Hsl1-Hsl7 Complex in the Presence of Osmotic Stress

Formation of the Hsl1-Hsl7 complex is necessary for Swe1 degradation [14]. This complex is located on the bud neck [14]. The regulation of Hsl1 controls the dynamics of the formation of the Hsl1-Hsl7 complex in untreated and treated (with osmotic stress) cells [58]. Therefore, we abstract the regulation of the Hsl1-Hsl7 complex by Hsl1. Activity of Hsl1 correlates with bud emergence and remains stable up to nuclear division. The localisation of Hsl1 and Hsl7 to the septin cortex takes place exactly after bud formation [58].

In the presence of osmotic stress, Hog1PP phosphorylates Hsl1, delocalises Hsl7 from the bud neck [17] and hinders the formation of Hsl1-Hsl7 complex. As a consequence, Swe1 is accumulated and the G2-to-M transition is blocked. We model the formation of the Hsl1-Hsl7 complex in the presence of Hog1PP activity by:

$$v_{phsl17} = \frac{kk_{hsl17}[BUD]}{1 + \left(\frac{[Hog1PP]}{J_{hsl1}}\right)^{n_{hsl1d}}}, \qquad (3.9)$$

where, BUD denotes a mathematical function that describes the observation of bud formation and depends on the levels of Cln2, Cln3 and Clb5 (Eq. A.36 of Appendix A) [13].

3.3 Model Description

Hsl1 is stable until nuclear division. Hsl1, like Clb2, is a substrate of APC (Anaphase-Promoting Complex) [10, 52]. Therefore, we assume that Hsl1 degrades with: (i) an APC dependent mechanism similar to Clb2; and (ii) also a natural self-degradation. Hence, the degradation of Hsl1-Hsl7 is modelled by:

$$v_{dhsl17} = kkk_{hsl17} V_{db2}[Hsl1\text{-}Hsl7] + kkd_{hsl17}[Hsl1\text{-}Hsl7], \qquad (3.10)$$

where $V_{db2} = kk_{db2} + kkk_{db2}[Cdh1] + k_{db2p}[Cdc20_A]$ represents the APC dependent degradation of the Hsl1-Hsl7 complex and the parameter kkd_{hsl17} captures the rate of self-degradation.

As a result, considering the Eqs. 3.9 and 3.10, the regulation of Hsl1-Hsl7 complex is described by:

$$\frac{d[Hsl1Hsl7]}{dt} = v_{phsl17} - v_{dhsl17} \qquad (3.11)$$

$$= \frac{kk_{hsl17}[BUD]}{1 + \left(\frac{[Hog1PP]}{J_{hsl1}}\right)^{n_{hsl1d}}}$$

$$- kkk_{hsl17} V_{db2}[Hsl1\text{-}Hsl7] - kkd_{hsl17}[Hsl1\text{-}Hsl7].$$

3.3.4 Regulation of Mih1

The protein phosphatase Mih1 reverses the phosphorylation of Cdc28-Clb2 by Swe1 [50]. We assume that there is a positive feedback between Cdc28-Clb2 and Mih1 [15]. This assumption is based on the reported positive feedback between Cdc25 and M-phase promoting factor (MPF) in frog extracts [28]; the homologues of Mih1 and Cdc28-Clb2 in *S. cerevisiae*. We use a Hill function to describe Mih1 regulation; activation by Clb2 and also self-inhibition.

$$\frac{d[Mih1]}{dt} = \frac{Va_{mih}[Clb2]([Mih1_T] - [Mih1])}{Ja_{mih} + [Mih1_T] - [Mih1]} - \frac{Vi_{mih}[Mih1]}{Ji_{mih} + [Mih1]}, \qquad (3.12)$$

where $[Mih1_T]$, the total concentration of Mih1, is constant.

3.3.5 Regulation of Cdc28-Clb2 in the Presence of Osmotic Stress

The pair of mitotic cyclins Clb1 and Clb2 are represented by Clb2 in the model. Clb2 is crucial for successful mitosis and its mutation causes G2 arrest [57]. Several mechanisms regulates the activity of Cdc28-Clb2 in the absence of osmotic stress. The activity of Hog1PP mediates the activity of Cdc28-Clb2 by: direct downregulation of *CLB2* transcription; and also accumulation of Swe1. The protein kinase Swe1 inhibits Cdc28-Clb2 by tyrosine phosphorylation of Cdc28. Consequently, presence of osmotic stress blocks the G2-to-M transition [17]. We consider the interaction of Clb2 with Swe1 and also Hog1PP in the model. The model for regulation of Cdc28-Clb2 consists of several parts:

(i) The availability of Clb2 is transcriptionally controlled (as soon as Clb2 is synthesised, it binds to Cdc28. This tethering is fast; therefore we do not distinguish between Clb2 and Cdc28-Clb2 in the model). Cdc28-Clb2 activates Mcm1, which is the transcription factor of *CLB2*, thereby establishing a positive feedback loop [3]. Also the presence of an osmotic stress (activity of Hog1PP) downregulates the transcription of the M phase cyclin (Clb2). Hence, the transcriptional production of Clb2 is described by:

$$v_{pClb2} = \frac{(kk_{sb2} + kkk_{sb2}[Mcm1])\,[mass]}{1 + (k_{dHog1Clb2}[Hog1PP])^{nHog1Clb2}},\tag{3.13}$$

where the parameter kk_{sb2} captures the basal transcription of *CLB2*, and kkk_{sb2} represents the induced expression of *CLB2* by Mcm1. Activity of Hog1PP also changes the transcription of *CLB2* (see Sect. 3.3.7). Note that we use Goldbeter-Koshland function to model the Mcm1 activity, Eq. A.79 of Appendix A.

(ii) The stoichiometric inhibitors, Sic1 and Cdc6, bind to Cdc28-Clb2 and inhibit its activity. Furthermore, when Sic1 or Cdc6 is in a complex with Cdc28-Clb2, the former can be phosphorylated. The phosphorylated Sic1 or Cdc6 degrades and releases Cdc28-Clb2 (see Fig. 2.6 of Chap. 2). Also natural degradation of Sic1 or Cdc6 frees Cdc28-Clb2 from the corresponding complexes (Cdc28-Clb2-Sic1, Cdc28-Clb2-Cdc6). We describe this tethering and releasing by the following equations:

$$v_{dC2} = k_{dib2}[C2],\tag{3.14}$$
$$v_{dC2P} = k_{d3c1}[C2P],$$
$$v_{dF2} = k_{dif2}[F2],$$
$$v_{dF2P} = k_{d3f6}[F2P],$$
$$v_{dC2h} = k_{hdib2}[C2h],$$
$$v_{dC2hP} = k_{hd3c1}[C2hP],$$
$$v_{pC2} = k_{asb2}[Sic1][Clb2],$$
$$v_{pF2} = k_{asf2}[Cdc6][Clb2],$$
$$v_{pC2h} = k_{hasb2}[Sic1h][Clb2],$$

where v_{dC2} and v_{dC2P} represent the release of Cdc28-Clb2 from the Cdc28-Clb2-Sic1 complex (denoted by C2) and the Cdc28-Clb2-Sic1P complex (C2P), respectively. Likewise, v_{dF2} and v_{dF2P} describe the unbinding of Cdc28-Clb2 from the Cdc28-Clb2-Cdc6 complex (F2), and from the Cdc28-Clb2-Cdc6P complex (F2P), respectively. Finally, v_{dC2h} and v_{dC2hP} illustrate the disassociation of Cdc28-Clb2 from the Cdc28-Clb2-Sic1h complex (C2h) and from the Cdc28-Clb2-Sic1hP complex (C2hP), respectively. Note that we distinguish between Sic1 which is phospho-

3.3 Model Description

Table 3.1 Table of abbreviation for different complexes of Cdc28-Clb2

Abbreviation	Complex	Abbreviation	Complex
C2	Cdc28-Clb2-Sic1	C2P	Cdc28-Clb2-Sic1P
F2	Cdc28-Clb2-Cdc6	F2P	Cdc28-Clb2-Cdc6P
C2h	Cdc28-Clb2-Sic1h	C2hP	Cdc28-Clb2-Sic1hP
PF2	Cdc28P-Clb2-Cdc6	PF2P	Cdc28P-Clb2-Cdc6P
PTrim	Cdc28P-Clb2-Sic1	PTrimP	Cdc28P-Clb2-Sic1P
PTrimh	Cdc28P-Clb2-Sic1h	PTrimhP	Cdc28P-Clb2-Sic1hP
C5	Cdc28-Clb5-Sic1	C5P	Cdc28-Clb5-Sic1P
C5h	Cdc28-Clb5-Sic1h	C5hP	Cdc28-Clb5-Sic1hP

rylated by CDK (Sic1P) and the one which is phosphorylated by Hog1PP (Sic1h) (see Sect. 3.3.8). Table 3.1 summarises the names that we used for different complexes of Cdc28-Clb2.

(iii) Swe1 inhibits Cdc28-Clb2 activity by tyrosine phosphorylation of Cdc28 [9, 50]. We assume that all three stable forms of the kinase Swe1 (Swe1, Swe1M and Swe1P; see Sect. 3.3.2 for details of their regulation) can inactivate Cdc28-Clb2. This phosphorylation is described by:

$$v_{iSwe1Clb2} = K_{Swe1}[Clb2], \tag{3.15}$$

where $K_{Swe1} = kk_{swe}[Swe1] + kkk_{swe}[Swe1M] + kkkk_{swe}[Swe1P]$.

(iv) Entry into mitosis requires Cdc28-Clb2 to be active. Mih1 reverses the phosphorylation of Cdc28 which is caused by Swe1 [50]:

$$v_{aMih1Clb2} = K_{Mih1}[PClb2], \tag{3.16}$$

where $K_{Mih1} = kk_{mih1}[Mih1] + kkk_{mih1}([Mih1_T] - [Mih1])$ and PClb2 represents the complex Cdc28-Clb2 in which Cdc28 is phosphorylated by Swe1 (Cdc28P-Clb2). The regulation of PClb2 is modelled by Eq. A.42 of Appendix A.

(v) Degradation of Clb2 is performed by two major APC (Anaphase-Promoting Complex) subunits: Cdc20 and Cdh1. During anaphase, Cdh1 is absent and Clb2 is primarily degraded by Cdc20. Cdc20 also triggers the pathway that results in the activation of Cdh1 [7, 62]. The remaining fraction of the Clb2 is degraded by Cdh1 during exit from mitosis [7]. The equation for this regulation is:

$$v_{dClb2} = V_{db2}[Clb2], \tag{3.17}$$

where $V_{db2} = kk_{db2} + kkk_{db2}[Cdh1] + k_{db2p}[Cdc20_A]$ represents the APC dependent degradation of Clb2.

Hence, the regulation of Cdc28-Clb2 can be summarised by the following equation:

$$\frac{d[Clb2]}{dt} = v_{pClb2} + v_{dC2} + v_{dC2P} + v_{dF2} + v_{dF2P} + v_{dC2h} + v_{dC2hP} \quad (3.18)$$
$$- v_{pC2} - v_{pF2} - v_{pC2h} + v_{iMih1Clb2} - v_{iSwe1Clb2} - v_{dClb2}$$
$$= \frac{(kk_{sb2} + kkk_{sb2}[Mcm1])\,[mass]}{1 + (k_{dHog1Clb2}[Hog1PP])^{n_{Hog1Clb2}}}$$
$$+ (k_{dib2}[C2] + k_{d3c1}[C2P]) + (k_{hdib2}[C2h] + k_{hd3c1}[C2hP])$$
$$+ (k_{dif2}[F2] + k_{d3f6}[F2P])$$
$$- (V_{db2} + k_{asb2}[Sic1] + k_{asf2}[Cdc6] + k_{hasb2}[Sic1h])\,[Clb2]$$
$$+ (K_{Mih1}[PClb2] - K_{Swe1}[Clb2]).$$

Moreover, we assume that Swe1 can also phosphorylate Cdc28 in the complexes of C2, C2P, C2h and C2hP (see Fig. 3.5 and Table 3.1 for abbreviation). We denote the complexes Cdc28P-Clb2-Sic1, Cdc28P-Clb2-Sic1P, Cdc28P-Clb2-Sic1h and Cdc28P-Clb2-Sic1Ph by PTrim, PTrimP, PTrimh and PTrimhP, respectively. Figs. 3.2b and 3.5 describe how these complexes relate to C2, C2P, C2h and C2hP. The regulation of PTrim, PTrimP, PTrimh and PTrimhP are modelled by Eqs. A.43, A.45,

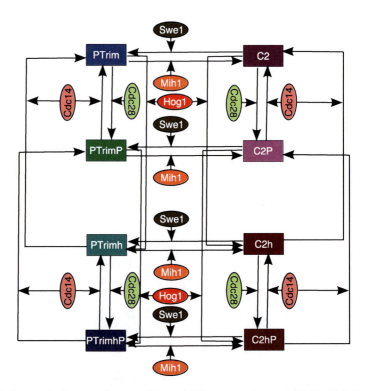

Fig. 3.5 Schematic diagram of the regulation of different complexes of Cdc28-Clb2. For details of interactions see text. The abbreviation used in this figure are defined in Table 3.1

3.3 Model Description 41

A.53 and A.54 of Appendix A, respectively. Likewise, we introduce the Cdc28P-Clb2-Cdc6 and Cdc28P-Clb2-Cdc6P complexes when Swe1 phosphorylates Cdc28 in the F2 and F2P complexes (see Fig. 3.2c). The Cdc28P-Clb2-Cdc6 and Cdc28P-Clb2-Cdc6P are denoted by PF2 and PF2P, respectively. Equations. A.46 and A.47 of Appendix A model their regulations.

3.3.6 Cell Growth

Osmotic stress arrests cells at different stages of their cycle; a pure exponential model for cell growth – as used by Chen et al. [13] for untreated cell – causes cells to reach unrealistic size in this condition. We therefore model the cell growth, represents by cell mass, with a logistic function which limits the cell growth in contrast to exponential function (Eq. 3.19). The cell volume is proportional to cell mass [12, 13]. Upon osmotic stress, cell volume decreases [32]. Hence, cell growth is slower compared with an untreated cell [1]. Therefore, the model which describes the cell growth in the presence of osmotic stress is:

$$\frac{d[mass]}{dt} = \frac{k_g[mass]\left(M_{mass,max} - [mass]\right)}{1 + k_{dhm}[Hog1PP]}, \tag{3.19}$$

where k_g is the growth rate in untreated conditions; $M_{mass,max}$ is the maximum reported mass for the cell; and k_{dhm} represents the influence of Hog1PP on the cell growth.

3.3.7 Influence of Hog1PP on the Cyclins Transcription

As we mention in Chap. 2, *S. cerevisiae* has 9 cyclins: the three G1 cyclins (Cln1, Cln2, Cln3) and the six B-type cyclins (Clb1 to Clb6) [39]. Cln3 is responsible for cell growth and the activation of Cln1 and Cln2 [61]. Cln1 and Cln2 act similarly [61], therefore they are lumped together as Cln2 in the model. The major functions of Clb5 and Clb6, which is the initiation of the DNA synthesis, is identical [6]. Hence, the pair of S phase cyclins is represented by Clb5 in our model. The set of mitotic cyclins, Clb1 and Clb2, is represented by Clb2 in this work. The remaining B-cyclins, Clb3 and Clb4, have a redundant role in initiating the S phase and can play a role during spindle formation, which can be substituted by Clb2 [39]. Therefore they are not distinguished from Clb2 in the model.

We assume that each cyclin (Cln3, Cln2, Clb5 and Clb2) is synthesised at a rate proportional to cell mass and then accumulates in the nucleus [12, 13] (Eqs. A.2, A.3, A.4 of Appendix A). Hog1PP is also located in the nucleus [46] and abruptly alters the transcription rate of cyclins [8, 17, 24, 63]. We model these alternations

42

in the transcription rate of cyclins by a Hill function (see Eqs. A.2, A.3, A.4 of Appendix A for the model of cyclins regulation).

3.3.8 Regulation of Sic1 Under Osmotic Stress

The activity of Sic1 in the G1 phase controls the G1-to-S transition [48]. Several mechanisms regulate Sic1:

(i) The availability of Sic1 is transcriptionally controlled. Swi5 (transcription factor) transcribes *SIC1* during the M-to-G1 transition and the G1 phase [22].

$$v_{pSic1} = kk_{sc1} + kkk_{sc1}[Swi5], \tag{3.20}$$

where kk_{sc1} is the basal production of Sic1, and kkk_{sc1} represents the rate of transcriptional synthesis of *SIC1* by Swi5. Regulation of Swi5 is modelled by Eq. A.19 of Appendix A.

(ii) Sic1 binds to the B-type cyclin complexes (Cdc28-Clb5 and Cdc28-Clb2) to inhibit their activity during the G1 phase. We model these associations by mass-action kinetics:

$$v_{pC5} = k_{asb5}[Clb5][Sic1], \tag{3.21}$$
$$v_{pC2} = k_{asb2}[Clb2][Sic1],$$

where k_{asb5} and k_{asb2} are the rates of Sic1 binding to Cdc28-Clb5 and Cdc28-Clb2 to build C5 and C2 complexes, respectively.

We also assume that Sic1 binds to Cdc28P-Clb2 (PClb2), and use mass-action kinetics to model this reaction:

$$v_{pPTrim} = k_{pasb2}[PClb2][Sic1], \tag{3.22}$$

where k_{pasb2} captures the association rate of Sic1 to PClb2.

(iii) To enable the G1-to-S transition, Sic1 has to be inactivated. Cdc28 phosphorylates and consequently inactivates Sic1 [48]. We model the phosphorylation of Sic1 by Cdc28 using mass-action kinetics; the total concentration of Cdc28 is proportional to its concentration in complexes with cyclins:

$$v_{pSic1P} = V_{kpc1}[Sic1], \tag{3.23}$$

where V_{kpc1} is described by Eq. A.84 of Appendix A. V_{kpc1} is a function of the total Cdc28 concentration (represented by cyclin concentration in the model) and the total concentration of Sic1.

(iv) During the M phase, active APC (Anaphase-Promoting Complex) complexes degrade the B-type cyclins (Clb5, Clb2). This degradation releases Sic1 from the C5

3.3 Model Description 43

(Cdc28-Clb5-Sic1) and C2 (Cdc28-Clb2-Sic1) complexes in addition to spontaneous disassociation of Sic1 from these two complexes. We model the freeing of Sic1 from C2 and C5 complexes (see Table 3.1 for abbreviations) by the following equations:

$$v_{aSic1C5} = (V_{db5} + k_{dib5}) [C5],$$ (3.24)
$$v_{aSic1C2} = (V_{db2} + k_{dib2}) [C2],$$

where $V_{db5} = kk_{db5} + kkk_{db5}[Cdc20_A]$ and $V_{db2} = kk_{db2} + kkk_{db2}[Cdh1] + k_{db2p}[Cdc20_A]$ capture the dynamics of APC dependent release of Sic1 from C5 and C2, and k_{dib5} and k_{dib2} are the disassociation rates of Sic1 from C5 and C2, respectively.

We assume that Sic1 is released from PTrim and C2 (see Table 3.1 for abbreviation) with similar mechanism. The following equation describes our model for freeing Sic1 from PTrim:

$$v_{aSic1PTrim} = V_{pdb2}[PTrim] + k_{pdib2}[PTrim],$$ (3.25)

where $V_{pdb2} = V_{db2}$ represents the APC dependent degradation of Clb2 and k_{pdib2} is the disassociation rate of Sic1 from PTrim.

(v) Protein phosphatase Cdc14 is required for exit from mitosis. Cdc14 dephosphorylates Sic1P. We model this dephosphorylation by mass-action kinetics. Regulation of Cdc14 is described by Eq. A.29 of Appendix A.

$$v_{pSic1Sic1P} = k_{ppc1}[Cdc14][Sic1P],$$ (3.26)

where k_{ppc1} is the dephosphorylation rate of Sic1P by Cdc14.

(vi) Hog1PP directly phosphorylates Sic1 [24]. This phosphorylation reduces the binding of Sic1 to the SCF (Skp1-cullin-F) complex and thus hinders efficient Sic1 degradation [24]. Phosphorylation of Sic1 by Hog1PP occurs at its Thr173 site, which is different from the Cdc28-dependent phosphorylation site. Therefore, we distinguish between the Sic1 which is phosphorylated by Cdc28 (Sic1P) and the one which is phosphorylated by Hog1PP (Sic1h). Sic1h is more stable compared to Sic1P [24]. We use a Hill function to model the phosphorylation of Sic1 by Hog1PP:

$$v_{iSic1Hog1PP} = \frac{kk_{ash}[Sic1][Hog1PP]^{nhsic1}}{kkk_{ash1}^{nhsic1} + [Hog1PP]^{nhsic1}},$$ (3.27)

where k_{ash1} corresponds to maximal level of Sic1h, and kkk_{ash1} represents the concentration of Hog1PP needed to phosphorylate Sic1.

We assume that Cdc14 is the phosphatase that reverses the phosphorylation of Sic1 by Hog1PP (see Figs. 3.2a, b and 3.5). This is a reasonable assumption, since Cdc14 has many substrates in a cell and has been reported to dephosphorylate many Cdc28-Clb substrates [55]. We use mass-action kinetics to model this dephosphorylation:

$$v_{pSic1Sic1h} = k_{h1ppc1}[Cdc14][Sic1h],$$ (3.28)

where k_{h1ppc1} is the dephosphorylation rate of Sic1h by Cdc14.

By similar arguments we model the regulation of Sic1P and Sic1h. We also assume that simultaneous phosphorylation of Sic1 by Cdc28 and Hog1PP is possible, the product of which is denoted by Sic1hP (see Table 3.1).

$$
\begin{aligned}
\frac{d[Sic1]}{dt} &= v_{pSic1} + v_{aSic1C5} + v_{aSic1C2} \\
&\quad + v_{aSic1PTrim} + v_{pSic1Sic1P} \\
&\quad + v_{pSic1Sic1h} - v_{pC5} - v_{pC2} - v_{pPTrim} \\
&\quad - v_{pSic1P} - v_{iSic1Hog1PP} \\
&= (kk_{sc1} + kkk_{sc1}[Swi5]) + (V_{db5} + k_{dib5})[C5] \\
&\quad + (V_{db2} + k_{dib2})[C2] + \left(V_{pdb2} + k_{pdib2}\right)[PTrim] \\
&\quad + \left(k_{ppc1}[Sic1P] + k_{h1ppc1}[Sic1h]\right)[Cdc14] \\
&\quad - \left(k_{asb2}[Clb2] + k_{asb5}[Clb5] + k_{pasb2}[PClb2] + V_{kpc1}\right)[Sic1] \\
&\quad - \frac{kk_{ash}[Sic1][Hog1PP]^{n_{hsic1}}}{kkk_{ash1}^{n_{hsic1}} + [Hog1PP]^{n_{hsic1}}}.
\end{aligned}
\tag{3.29}
$$

$$
\begin{aligned}
\frac{d[Sic1P]}{dt} &= V_{kpc1}[Sic1] + V_{db2}[C2P] + V_{db5}[C5P] \\
&\quad + V_{pdb2}[PTrimP] + k_{h1ppc1}[Sic1hP][Cdc14] \\
&\quad - \left(k_{ppc1}[Cdc14] + k_{d3c1}\right)[Sic1P] \\
&\quad - \frac{kk_{ash1}[Sic1P][Hog1PP]^{n_{hsic1}}}{kkk_{ash1}^{n_{hsic1}} + [Hog1PP]^{n_{hsic1}}}.
\end{aligned}
\tag{3.30}
$$

$$
\begin{aligned}
\frac{d[Sic1h]}{dt} &= \frac{kk_{ash}[Sic1][Hog1PP]^{n_{hsic1}}}{kkk_{ash1}^{n_{hsic1}} + [Hog1PP]^{n_{hsic1}}} \\
&\quad + V_{hdb2}[C2h] + k_{hdib2}[C2h] \\
&\quad + V_{hdb5}[C5h] + k_{hdib5}[C5h] + k_{hppc1}[Cdc14][Sic1hP] \\
&\quad - \left(k_{hasb2}[Clb2] + k_{hasb5}[Clb5] + V_{hkpc1}\right)[Sic1h] \\
&\quad + - \left(k_{hpasb2}[PClb2] + k_{h1ppc1}[Cdc14]\right)[Sic1h] \\
&\quad + \left(k_{hpdib2} + V_{hpdb2}\right)[PTrimh].
\end{aligned}
\tag{3.31}
$$

3.3 Model Description

$$\frac{d[Sic1hP]}{dt} = V_{hkpc1}[Sic1h] - \left(k_{hppc1}[Cdc14] + k_{hd3c1}\right)[Sic1hP] \quad (3.32)$$
$$+ V_{hdb2}[C2hP] + V_{hdb5}[C5hP] + V_{hpdb2}[PTrimhP]$$
$$- k_{h1ppc1}[Sic1hP][Cdc14] + \frac{kk_{ash1}[Sic1P][Hog1PP]^{n_{hsic1}}}{kkk_{ash1}^{n_{hsic1}}[Hog1PP]^{n_{hsic1}}}.$$

3.4 Parameter Estimation

The set of parameters used in our simulation is presented in Appendix A. These parameters are chosen from the literature, when available [12, 13, 15, 18, 21, 23–25, 56, 64, 65]. The parameters used in the modelling of the morphogenesis checkpoint are adapted from Ciliberto et al. [15]. The remaining parameters are determined by comparing the simulation outputs of the mathematical model (e.g. delay duration) to the experimental observations in wild-type and mutated yeast cells [17, 24, 63], as explained below.

The parameters involved in the interaction of Hog1PP and the cell cycle components are chosen based on the known delay duration of wild-type cells and different mutated cells under different doses of osmotic stress. Note that the delay duration is the difference between length of untreated and treated cell cycle as defined by Eq. 3.33. Presence of 0.4 M NaCl activates the Hog1PP as shown in Fig. 3.6.

The simulation results for the 0.4 M NaCl dose reproduce the delay duration for wild-type and several mutated cells as reported by Escote et al., Clotet 2006 et al. and Yaakov 2009 et al. [17, 24, 63]. Figure 3.7 shows the delay duration according to our model for wild-type cell which is exposed to 0.4 M NaCl. Moreover, the model reproduces the observed delay for the $sic1\Delta$ cell [24, 63] and $swe1\Delta$ cell [17, 63] (see Figs. 3.8 and 3.9).

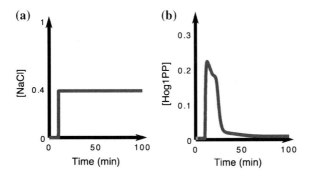

Fig. 3.6 Activation profile of Hog1PP in response to a single step of 0.4 M NaCl

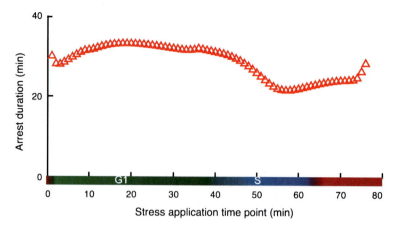

Fig. 3.7 Simulation results for the delay duration when the cell is exposed to 0.4 M NaCl. These results are comparable to the experimental measurements [17, 24, 63]

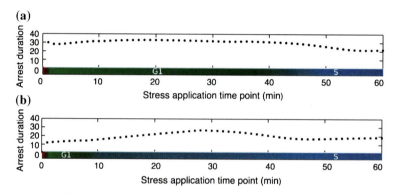

Fig. 3.8 Simulation results for the arrest duration for (**a**) wild-type and (**b**) $sic1\Delta$ cells in response to the presence of 0.4 M NaCl. Note that knock-out mutation of Sic1 from the cell changes the length of the G1 phase of the cell cycle; the length of the G1 phase of $sic1\Delta$ cell is shorter compare with the wild-type cell. The model reproduces the observed delay for wild-type and $sic1\Delta$ cells [24, 63]

3.5 Model Predictions

So far we have translated the known biological mechanisms for the interaction between the osmotic stress signalling pathway and the cell cycle network of *S. cerevisiae* [2, 8, 17, 24, 63] into a comprehensive mathematical model. These translations lead to a model which predicts: the response of the cell cycle during the S and M phases to osmotic stress; the effect of various doses of osmotic stress on the entire cell cycle; and the emergent dynamics as a result of the cell cycle response to osmotic tress. The key predictions of model are: (i) osmotic stress delays the G1-to-S and G2-to-M transitions in a dose dependent manner, (ii) cells stressed at late G2/M

3.5 Model Predictions   47

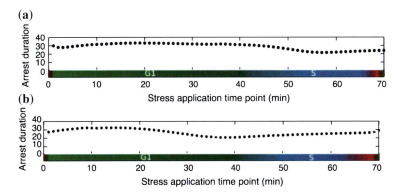

Fig. 3.9 Simulation results for the arrest duration for (**a**) wild-type and (**b**) $swe1\Delta$ cells in response to the presence of 0.4 M NaCl. The model reproduces the observed delay for wild-type and $swe1\Delta$ cells [17, 63]

phase display dose-independent accelerated exit from mitosis and arrest in the next cell cycle, (iii) exposure of the cell to osmotic stress during the late S and early G2/M phase can induce DNA re-replication, (iv) the HOG MAPK network can take over the role of the MEN network during cell division. In this section, we present the predictions of the model and the experimental evidences which support them. Based on the model analysis, we discuss the molecular mechanisms which emerge as a consequence of cell cycle response to osmotic stress [45].

3.5.1 Osmotic Stress Delays the G1-to-S and G2-to-M Transitions

The osmostress-activated MAP kinase Hog1 modulates the activity of several components of the cell cycle network, to prevent cell cycle progression before proper adaptation to the osmotic stress (see Chap. 2 for details of these interactions). According to our model and in accordance with experimental observations; the osmotic stress induces cell cycle arrest depending on the cell cycle phase at the onset of stress.

To demonstrate this, we induce osmotic stresses at different points of the cell cycle model (Fig. 3.10 shows the activation profile of Hog1PP for various doses of stress) and calculate the arrest duration by:

$$\tau = T_{\text{stress}} - T_{\text{untreated}}, \tag{3.33}$$

where T_{stress} denotes the cell cycle duration under stress and $T_{\text{untreated}}$ is the cell cycle duration of untreated cells.

Figure 3.11 shows simulation results of the arrest duration; τ throughout the cell cycle for different doses of NaCl. Strikingly, there are two distinct types of cell cycle responses to NaCl depending on the timing of stress:

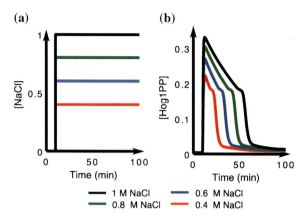

Fig. 3.10 Dose-dependent activation of Hog1PP. Different colours shows various doses of the stress, ranging from 0.4 M NaCl to 1 M NaCl. The higher doses of NaCl cause longer and stronger Hog1PP activity

(i) Before the transition to the M phase, τ is positive; and the cell cycle progression becomes slower (Fig. 3.11). There is some variability in the value of τ depending on the time point at which the stress is applied. For example, the cell cycle arrest reaches the minimum duration within the S phase and increases again at the beginning of the G2 phase.

(ii) Immediately after the beginning of the FINISH process,[1] instead of having a delay in the cell cycle, the progression of the cell cycle is accelerated; τ becomes negative. Cells exposed to osmotic stress at that time or later have an accelerated M-to-G1 transition and get arrested in the G1 phase of the new cell cycle. Remarkably, the transition in τ is very sharp. The stress induced arrest duration changes suddenly from being close to 40 min to approximately -35 min for 0.5 M NaCl (green circle in Fig. 3.11). The later the stress is applied after FINISH, the smaller becomes τ in absolute value and the cell cycle duration becomes closer to the cell cycle duration of the untreated cell.

Our model systematically investigates the influence of different suggested molecular mechanisms [2, 8, 17, 24, 63] on the induced delay across the different cell cycle phases in response to osmotic stress. During the G1 phase, the reported mechanisms responsible for the delay are; (i) the Sic1 stabilisation mediated by Hog1PP; and (ii) the transcription downregulation of *CLN2* and *CLN3* due to Hog1PP [8, 24].

To assess the importance of the first mechanism, we block the interaction of Sic1 with Hog1PP in the model. To implement this scenario experimentally the specific phosphorylation side of Sic1 (T173) should be blocked [24]. In this case τ strongly decreases, especially at the beginning of the G1 phase, compared with the respective arrest duration in the wild type (red cross in Fig. 3.12). At the end of G1 phase; as the

---

[1] The beginning of the FINISH process is defined as the time point at which Mcm1 reaches its maximal value [12].

3.5 Model Predictions

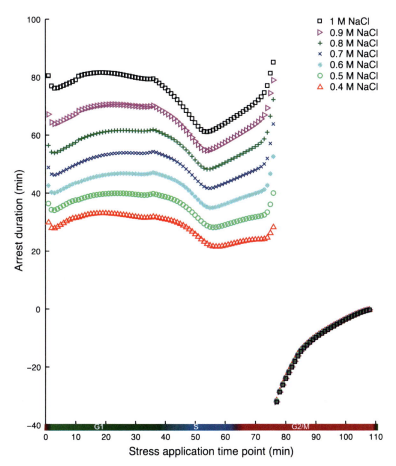

Fig. 3.11 Dose-dependent arrest duration following the imposition of osmotic stress at different stages of the cell cycle. The x-axis represents the time point of application of stress, whereas the y-axis illustrates the corresponding arrest duration. Different colours demonstrate various doses of the stress, ranging from 0.4 M NaCl to 1 M NaCl. During the G1 phase and the S phase, higher doses of stress cause longer cell cycle arrests, while the acceleration of the exit from mitosis cell is dose independent

level of Sic1 decreases in wild-type cell the role of Sic1 becomes less dominant in blocking the G1-to-S transition. In contrast, if we remove the influence of Hog1PP on the transcription of *CLN2* and *CLN3*, we observe a small difference in the delay τ compared with the wild-type cell (compare the blue circle with black triangular in Fig. 3.12). Therefore, according to the model Sic1 stabilisation by Hog1PP plays an essential role in the G1 phase arrest.

The regulation of Cdc28-Cln2 and Sic1 upon activation of Hog1PP during G1 is shown in Fig. 3.13b and c. The dynamical behaviour of the G1 components of our model is similar to the recently published model by Adrover et al. [1]

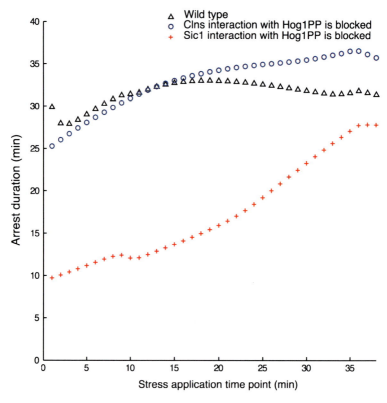

Fig. 3.12 Assessing the role of two key mechanisms responsible for the cell adaptation to osmotic stress during the G1 phase. The x-axis represents the time point of application of stress, whereas the y-axis illustrates the corresponding arrest duration. The cell cycle model is exposed to 0.4 M NaCl. Blocking the interaction of Sic1 with Hog1PP in the model, reduces the arrest duration significantly along the G1 phase (*red cross*)

(see Figs. 3.13a–c). That model studies the response of the G1 phase cell to osmotic stress and suggests that Hog1 delays bud morphogenesis and DNA replication by controlling the regulation of Clns and Clb5 respectively [1]. Figure 3.14 shows the Summary of the interactions of Hog1PP with cell cycle components according to our model. In stark contrast to the model by Adrover et al., our model predicts the activity profile of the Hog1PP targets when the different doses of stress are applied during the G1-to-S, G2-to-M and M-to-G1 transitions (see Figs. 3.13a–f, 3.15a–f). The model also investigates the role of different interactions in blocking the G2-to-M transition.

During the early S phase, the delayed accumulation of Cdc28-Clb5 is the suggested mechanisms for slow cell cycle progression [1, 63]. According to our model; first, the downregulation of *CLB5* transcription and accumulation of Sic1 by Hog1PP cause the Cdc28-Clb5 level to decrease initially. However, after Hog1PP returns to its basal level, this model predicts a further increase of Cdc28-Clb5 by the active

3.5 Model Predictions

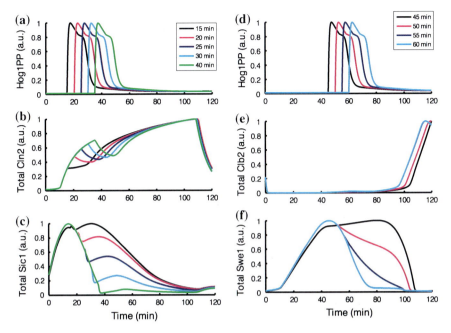

Fig. 3.13 Model prediction for the activity of Hog1PP and its target during the G1-t-S and G2-to-M transitions. (**a**) 0.4 M NaCl is applied at different time points during the G1 phase to study the influence of stress timing on the profile of Hog1PP. Activity of Hog1PP is independent of the timing of osmotic stress during the G1 phase. (**b**) Presence of 0.4 M NaCl at different time points during the G1 phase changes the activity profile of Cln2. Osmotic stress causes downregulation of Cln2, which is followed by an upregulation when Hog1PP is deactivated again. (**c**) Presence of 0.4 M NaCl at different time points during the G1 phase accumulates Sic1. Moving towards the S phase this accumulation is less pronounced. (**d**) 0.4 M NaCl is applied at different time points during the G2-to-M transition. The time profile of Hog1PP is shown. (**e**) The activation of Clb2 is delayed due to presence of Hog1PP. (**f**) Activity of Hog1PP causes the accumulation of Swe1. Moving towards the M phase this accumulation is less pronounced. The concentration of the components have been normalised for illustration purpose; a.u., arbitrary units

SBF/MBF transcription factors (compare the activity of Cdc28-Clb5 in Fig. 3.16b with a). The SBF/MBF transcription factors, in turn, remain high for an extended period of time due to Hog1PP mediated stabilisation of Swe1 (see Fig. 3.14). The accumulation of Swe1 prevents the increase of Clb2 activity, which is the main inhibitor of SBF/MBF. Moreover, SBF/MBF activates the transcription of Swe1, establishing a positive feedback mechanism (see Fig. 3.14 for the positive feedback between SBF/MBF and Swe1 via Cdc28-Clb2), causing the maximum level of Swe1 to be higher than in the unstressed cell cycle (see Figs. 3.13f and 3.15f).

Only when the level of the Hsl1-Hsl7 complex increases, the feedback mechanism becomes less efficient, enabling the transition to the late S phase. The stress-induced delay during early S phase shows a steady decrease, as shown in Fig. 3.11, due to the higher level of the Hsl1-Hsl7 complex as we move towards late S phase, making the

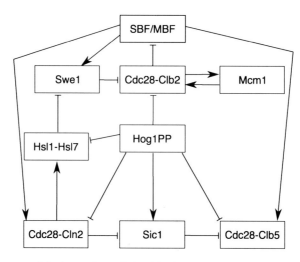

Fig. 3.14 Summary of the interactions of Hog1PP with cell cycle components. The investigation with the model suggests this simple network which presents the important interactions of the cell cycle in response to the activity of Hog1PP. Hog1PP phosphorylates Sic1, which accumulates Sic1, and blocks the G1-to-S transition. Also, Hog1PP phosphorylates Hsl1 and prevents the formation of the Hsl1-Hsl7 complex. This leads to the accumulation of Swe1, due to the double negative link between Hog1PP and Swe1. This hinders the G2-to-M transition. Transcriptional downregulation of *CLN2* by Hog1PP influences the cell cycle progression in two different ways: first, it causes Sic1 to be less phosphorylated and more active. This inhibits the cell cycle transition to the S phase. Second, it influences indirectly the timing of the formation of the Hsl1-Hsl7 complex, which is important for the transition to the G2/M phase. Moreover, the downregulation of Cdc28-Clb2 by Hog1PP can cause SBF/MBF to be active for an extended duration, and therefore a cell cycle arrest before the G2/M transition

positive feedback loop between SBF/MBF and Swe1 less effective (see Fig. 3.14). Consequently, SBF/MBF is active for a shorter period of time and, therefore, the delay decreases.

The role of Swe1 in the S phase delay caused by osmotic stress was tested experimentally by Yaakov et al., where it was found that a strain lacking Swe1 has almost the same delay during S phase as the wild type [63]. To validate our model, we perform a simulation removing Swe1, and in accordance with the experimental results, we obtain a delay during the S phase that differs only by approximately 15 min compared to the wild type (compare Fig. 3.16c with b).

The role of Sic1 in the S phase delay was also experimentally tested [63]. It was shown by Yaakov et al. that Sic1 has almost no influence on the delay of the cell cycle progression during the S phase [63]. Our model also reproduces this result (compare Fig. 3.16d with b). Therefore, our model suggests that the S phase delay due to osmotic stress cannot be attributed to a single component but rather it emerges as the result of the interaction among the many components of the cell cycle network.

3.5 Model Predictions

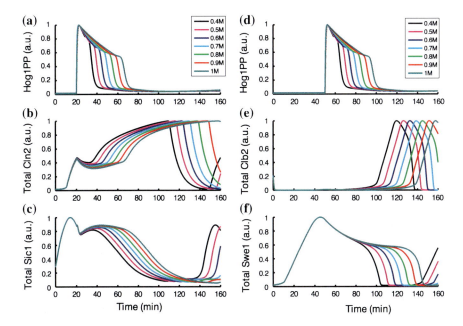

Fig. 3.15 Model prediction for the influence of the dose of NaCl on the activity of Hog1PP and its cell cycle components during G1-to-S and G2-to-M transition. The concentrations of the components have been normalised for illustration purpose. Different concentrations of NaCl can change the activity duration of Hog1PP. (**a**) Different strengths of stress are applied at a fixed time point which is located near START. (**b**) Hog1PP stabilises the level of Cln2 for a transient duration which is linearly proportional to the dose of stress. The transient delay duration increases linearly with the dose of stress. (**c**) Higher doses of NaCl cause stronger accumulation of Sic1. (**d**) Different concentrations of NaCl are applied to a fixed time point, which is located before G2-to-M transition. (**e**) A higher dose of stress will delay the upregulation of Clb2. (**f**) Higher doses of NaCl cause stronger accumulation of Swe1. The relation between the dose of stress and delay duration is also linear in this interval

If the stress is applied during late S phase or beginning of G2/M phase, a striking phenomenon occurs: A second incidence of DNA replication before cell division will take place. We discuss this in Sect. 3.5.4.

3.5.2 Osmotic Stress Causes Accelerated Exit from Mitosis

In contrast to the cases discussed above, if the osmotic stress is applied after a very precise time point in the G2/M phase (discontinuity in arrest duration in Fig. 3.11), the cell experiences an accelerated exit from mitosis. The time point at which this dramatic change occurs is determined by the point at which the level of Mcm1, the transcription factor of *CLB2*, reaches its maximum. This time point coincides with the initiation of the FINISH process [12]. Hence, if the osmotic stress is applied at that

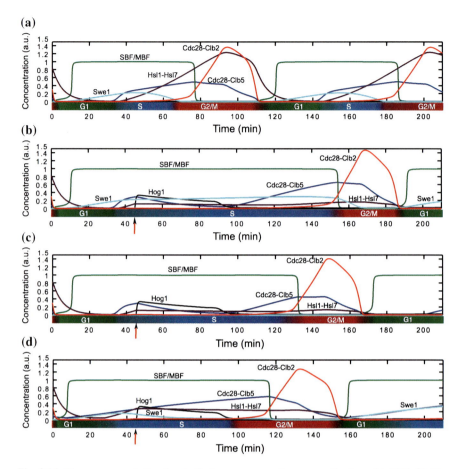

Fig. 3.16 Time course activity of early S phase components upon application of 1 M NaCl. The concentrations are rescaled for illustration purpose. The time point of application of stress is shown by red arrow. (**a**) Wild type untreated cell, (**b**) 1 M NaCl applied during early S phase (at t = 45 min) to a wild type cell causes the cell cycle to last about 76 min longer compared to the wild type untreated cell. (**c**) 1 M NaCl applied to a *swe1*Δ cell; in this case the cell cycle duration is 62 min longer than in an untreated *swe1*Δ cell. (**d**) The deletion of Sic1 does not cancel the delay caused by Hog1PP activity. 1 M NaCl applied to *sic1*Δ cell prolongs the cell cycle around 52 min compared to *sic1*Δ untreated cell

time point or later, our model predicts that the level of Cdc28-Clb2 starts to decrease immediately mainly due to two mechanisms: (i) inhibition of Cdc28-Clb2 activity by Sic1, the latter being stabilised by Hog1PP and, (ii) direct transcriptional inhibition of *CLB2* by Hog1PP. Hence, the point at which Cdc28-Clb2 starts decreasing occurs earlier than in the absence of stress (compare Fig. 3.17a with b). Moreover, the level of Sic1 starts increasing rapidly due to the presence of Hog1PP, and as a consequence, the exit from mitosis is significantly accelerated. Also, the later the stress is applied

3.5 Model Predictions

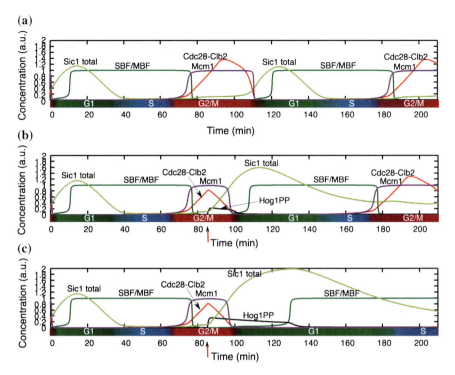

Fig. 3.17 Time course activity of M phase components upon application of 0.4 M and 1 M NaCl. The time point of application of stress is shown by *red arrow*. (**a**) Wild type untreated cell. (**b**) 0.4 M NaCl applied during G2/M phase to a wild type cell. Osmotic stress causes the cell to finish the current cell cycle very rapidly compared to untreated cell and gets arrested in the G1 phase of the next cell cycle. (**c**) 1 M NaCl applied during the G2/M phase. The time profile activity of the G2/M phase is the same for 0.4 M NaCl. This cell also experiences accelerated exit from mitosis and gets arrested in the G1 phase of the new cell cycle. In this case the G1 phase is longer compare to the cell treated with 0.4 M NaCl

during the M phase, the slower the acceleration becomes, since the exit from mitosis is further advanced (see Fig. 3.11).

Importantly, after the accelerated exit from mitosis, the cells get arrested in the next G1 phase. Depending on the dose of the stress, the delay can be carried over to the subsequent cell cycles (compare Fig. 3.17b with c). The effect of the stress dose is discussed in the next section.

3.5.3 Delays in the G1-to-S and G2-to-M Transitions are Dose Dependent, Whereas Acceleration of the M-to-G1 Transition is Dose Independent

Next we apply different stress doses, ranging from 0.4 M NaCl to 1 M NaCl, at different times along the entire cell cycle, following the approach described in the last section. Upon the onset of the stress, Hog1PP rises almost immediately, stays active for a time interval proportional to the stress dose, and then returns rapidly to its basal level (see Fig. 3.10). Notably, if the stress is applied before FINISH, the delay obtained increases approximately linearly with the stress dose (see Fig. 3.11), whereas if the stress is applied after that point, the acceleration does not depend on the stress dose (see Fig. 3.11).

According to the model, one of the key mechanisms responsible for this very different behaviour, is the regulation of the main transcription factors of the G1/S and G2/M phases, namely, SBF/MBF and Mcm1, respectively. The transcription complexes SBF/MBF are downregulated by Cdc28-Clb2, the latter being downregulated by Hog1PP (see Fig. 3.14). Hence, upon stress, SBF/MBF are activated for an extended time interval and the main G1 and S cyclins are stabilised until Hog1PP returns to its basal level. Hence, the delay in the cell cycle progression increases with the stress dose.

In contrast, Mcm1 is upregulated by Cdc28-Clb2, and upon stress, Cdc28-Clb2 is downregulated; as a consequence, the time interval during which Mcm1 is active, is reduced. Therefore, there is no stabilising influence on Cdc28-Clb2 upon stress, and its level decreases almost immediately after the onset of the stress, nearly independent of the stress dose (between 0.4 M-1 M NaCl). However, the next cell cycle will be delayed, since Hog1PP is still active; and the length of the delay in the next cell cycle depends on the stress doses (compare Fig. 3.17b with c).

3.5.4 Osmotic Stress at Late S or Early G2/M Phase Causes DNA Re-Replication

By simulating the application of osmotic stress in late S phase or early G2/M phase, the model predicts the initiation of a second incidence of DNA replication. This effect is more pronounced for higher stress doses. This prediction can be claimed based on the experimental evidence reported on the dual role of Cdc28-Clb5 [20, 42].

Cdc28-Clb5 first initiates the DNA replication at the onset of the S phase; and then blocks the assembly of the pre-replicative complex during the G2/M phase [20]. Blocking the assembly of the pre-replicative complex prevents DNA re-replication, thereby enabling stable propagation of genetic information [42]. Cdc28-Clb5 inhibits DNA re-replication by three overlapping mechanisms: first, Cdc28-Clb5 reduces Cdc6 levels through phosphorylation; second, it promotes the nuclear export of MCM proteins; and third, it phosphorylates ORC proteins [42]. All three mechanisms render

3.5 Model Predictions

the replication origins in the post-replicative state, so that the high level of Cdc28-Clb5 prevents *de novo* assembly of the pre-replicative complex during the G2/M phase [42].

In order to show that the activity of Cdc28-Clb5 is crucial to prevent DNA re-replication, Dahmann et al. mutated the *SIM* genes from cells; these mutation caused a second incidence of DNA replication without mitosis [20]. It was shown that mutated *SIM* genes lower the activity of Cdc28-Clb5, probably by a post-transcriptional mechanism [20]. To validate the key role of Cdc28-Clb5 activity in preventing DNA re-replication, Dahmann et al. overexpressed *CLB5* in the *SIM* mutant-G2-phase-arrested cells. Overexpression of *CLB5* inhibited DNA re-replication in those cells [20].

Moreover, in another experiment they inhibited Cdc28-Clb5 activity by inducing Sic1 expression during the G2/M phase; this inhibition led to the assembly of the pre-replicative complex [20]. Subsequent repression of Sic1 allowed recovery of Cdc28-Clb5 activity and, crucially, it triggered DNA re-replication [20]. Therefore, by lowering the activity of Cdc28-Clb5, the three DNA re-replication blocking mechanisms are rendered ineffective [20].

Interestingly, our model predicts that, by applying osmotic stress during late S or early G2/M phase, the activity of Cdc28-Clb5 is decreased. Just before applying the stress at that stage of the cell cycle, Cdc28-Clb5 has reached a high level and DNA replication is almost complete. If the osmotic stress is applied at that moment, initially Cdc28-Clb5 activity decreases; Hog1PP first downregulates *CLB5* transcription; and then stabilises Sic1, which also inhibits Cdc28-Clb5 activity. Furthermore, Hog1PP also downregulates *CLB2* expression and, as a consequence, the transcription factors SBF/MBF remain active for longer. Accordingly, the levels of Cdc6 and Cdc14 increase, leading to the assembly of the pre-replicative complex. Then, after Hog1PP returns to its basal level, Clb5 starts increasing again, leading to a second peak in Cdc28-Clb5 activity (compare Fig. 3.18a with b). This sequence of events therefore strongly suggests that a second incidence of DNA replication occurs before mitosis upon application of osmotic stress during late S or early G2/M phase. This novel prediction is in accordance with the experimental results mentioned above, which show that a decrease followed by an increase in Cdc28-Clb5 activity leads to DNA re-replication.

In order to further validate our model with the known measured data, we "numerically overexpressed" *CLB5* by simulating induction of *CLB5* transcription from the GAL1 promoter [13]. Note that this changes the length of the cell cycle phases as shown in Fig. 3.18c. In accordance with the experimental results, this overexpression inhibited DNA re-replication (see Fig. 3.18c). Moreover, to test the mechanism that we propose for DNA re-replication: namely, interaction of Hog1PP with Sic1, we blocked this interaction. This case led to inhibition of DNA re-replication (see Fig. 3.18d).

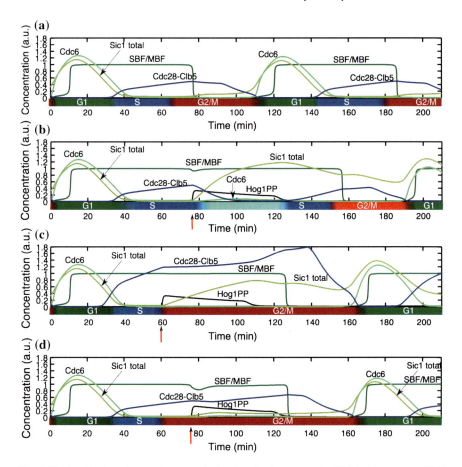

Fig. 3.18 Application of osmotic stress during late S phase or early G2/M phase causes DNA re-replication. The time point of application of stress is shown by *red arrow*. (**a**) Time course of activity of the cell cycle components for the wild type untreated cell. (**b**) 1 M NaCl applied at minute 76. Activity of Hog1PP causes downregulation of Cdc28-Clb5. Absence of Clb5 activity can lead to the assembly of the pre-replicative complex before cell division: Cdc6 level increases when Cdc28-Clb5 activity is reduced by Hog1PP. Then, after Hog1PP returns to its basal level, Clb5 starts increasing again. The downregulation, following by an upregulation of Cdc28-Clb5 can lead to DNA re-replication. (**c**) Overexpression of *CLB5*–by simulating induction of *CLB5* transcription from the GAL1 promoter – inhibits the DNA re-replication. (**d**) Blocking the interaction of Sic1 with Hog1PP also hinders the DNA re-replication in the presence of 1 M NaCl

3.5.5 Stabilisation of Sic1 by Hog1PP Drives the Mitotic Exit in MEN Mutant Cells

Our integrative cell cycle and osmotic stress model also provides a mechanistic explanation for the experimental results obtained by Reiser et al. regarding the response of *cdc15^{ts}* cells to osmotic stress [47].

3.5 Model Predictions

The protein kinase Cdc15 is one of the components of the Mitotic Exit Network (MEN) [5]. Cdc15 is active during the M phase and responsible – as well as other MEN components – for the final M-to-G1 transition. For this transition to occur, Cdc28-Clb2 has to be inactivated, which is partly mediated by the protein phosphatase Cdc14 [29, 40]. In *S. cerevisiae*, Cdc14 is localised in the nucleolus during most of the cell cycle, but it is released during late M phase [54, 60], thereby reversing the activity of Cdc28-Clb2. Cdc14 delocalisation from the nucleolus is mainly controlled by MEN [30, 35]. Hence, $cdc15^{ts}$ cells, as well as other viable MEN temperature-sensitive mutant cells, keep Cdc14 trapped in the nucleolus, and consequently the levels of Cdc28-Clb2 activity remain high [29]. This leads to $cdc15^{ts}$ cells being arrested in the M phase, unable to complete their cell cycles [27].

We first examine whether the model can reproduce the cell cycle arrest caused by mutation of Cdc15. Figure 3.19a shows the simulation result for a $cdc15^{ts}$ cell.[2] In accordance with the experimental results, the model shows that $cdc15^{ts}$ cells do not divide.

It was observed that the MEN mutant cells, at the non-permissive temperature, complete mitosis and cell division following imposition of osmotic stress [27, 47]. It was suggested that the measured increase in Cdc14 activity, induced by the HOG MAPK network, was responsible for the M-to-G1 transition of the cell in the perturbed environment [47]. The molecular mechanism behind this experimental observation, however, has remained unclear.

By simulating the response of $cdc15^{ts}$ cells to osmotic stress, we reproduce the reported experimental results. The simulations show that $cdc15^{ts}$ cells, as well as further temperature-sensitive MEN mutants, end mitosis and enter a new cell cycle after exposure to various doses of osmotic stress, as measured in the experiments (see Fig. 3.19b).

In order to identify the mechanism responsible for this response, we first tested the role of Cdc14 in cell division in the untreated, and treated with osmotic stress condition. According to the model, $cdc14^{ts}$ cells are arrested in M phase in the untreated condition. Presence of osmotic stress causes $cdc14^{ts}$ cells to go through cell division (see Fig. 3.19c). The same result was experimentally obtained by Grandin et al. [27]. This suggests that Cdc14 upregulation by Hog1PP is not the key mechanism for cell division of MEN mutants in the presence of osmotic condition.

We then tested the role of the downregulation of *CLB2* by Hog1PP in $cdc15^{ts}$ cells by blocking this interaction in silico and applying the stress in the M phase. In this case the cell could complete mitosis and progress to the next cell cycle (see Fig. 3.19d). Only by blocking the interaction between Hog1PP and Sic1, is possible to stop the cell from completing mitosis (see Fig. 3.19e).

Thus, the model reproduces the experimental observation regarding the response of MEN mutant cells to osmotic stress [27, 47]. Moreover, it suggests the key mechanism responsible for this observation: stabilisation of Sic1 by Hog1PP transfer the MEN mutant cells under osmotic stress to the G1 phase of the next cell cycle.

---

[2] $cdc15^{ts}$ cell is achieved by changing the value of kkk_{ppnet} (Table A.2 of Appendix A) from 3, to 0 [49].

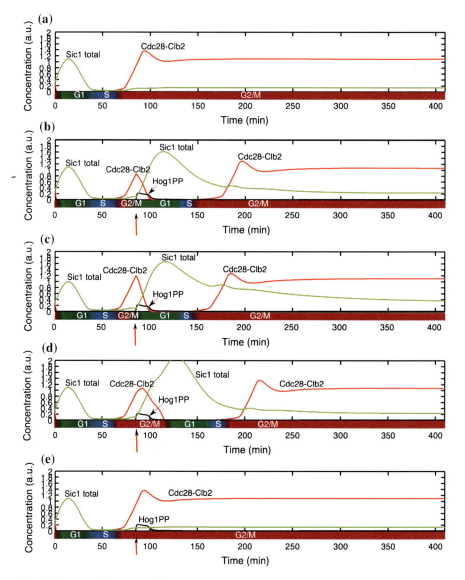

Fig. 3.19 The HOG MAPK network rescues the mitotic exit defect of MEN mutants. The time point of application of stress is shown by red arrow. (**a**) *cdc15^{ts}* cell is being arrested in M phase and cannot divide. (**b**) Application of 0.4 M NaCl stimulates the *cdc15^{ts}* cell to go through cell division. Note that the *cdc15^{ts}* cell upon osmotic stress is able to finish its current cell cycle but get arrested in the next G2/M phase. (**c**) The *cdc14^{ts}* cell can go through the cell division in the presence of 0.4 M NaCl. (**d**) Removing the interaction of Hog1PP with *CLB2* does not cancel the cell division of the *cdc15^{ts}* cell in the presence of osmotic stress. (**e**) The *cdc15^{ts}* cell, in which the interaction of Hog1PP with Sic1 is blocked, cannot finish its cell cycle and is arrested in M phase

3.6 Sensitivity Analysis of the Model

We study the influence of parameters on the model predictions throughout the cell cycle. Our focus is on the set of parameters which is involved in the response of the cell cycle to osmotic stress (shown in red in Table A.2 of Appendix A).

To investigate the sensitivity of the predicted delay duration to parameters we take the following approach:

(i) The set of parameters (shown in red in Table A.2 of Appendix A) is divided into four groups (Table 3.2). Each group consists of the parameters that are involved in the interactions between Hog1PP and the components of a specific stage of the cell cycle.

(ii) We then choose a time point in each of these groups for the application of 1 M NaCl: $t = 20$ min for G1 phase; $t = 50$ min for S phase; $t = 75$ min for DNA re-replication window; and $t = 90$ min for M phase.

(iii) For each group we generate 100 sets of randomly chosen parameters; logarithmically distributed in the interval from 0.1 to 10 times their value in Table A.2 of Appendix A. For example, group I in Table 3.2 consists of the 22 parameters, which involve in the interactions of Hog1PP with G1 components; Sic1, Cln2 and Cln3. 100 sets are chosen from logarithmically distributed random numbers in the interval from 0.1 to 10 times value of each of these 22 parameters in Table A.2 of Appendix A.

(iv) We repeat the simulation 100 times, for each of these four groups using the corresponding 100 sets of randomly generated parameters. The delay duration in response to the presence of 1 M NaCl for each of these simulation is then calculated using the Eq. 3.34:

$$\tau_{ri} = T_{i\text{stress}} - T_{\text{untreated}}, \tag{3.34}$$

where $T_{i\text{stress}}$ is the length of the cell cycle with the ith set of parameters when it is treated with 1 M NaCl; $T_{\text{untreated}}$ is the length of the untreated cell cycle with the set of parameters listed in Table A.2 of Appendix A; and τ_{ri} represents the delay duration for the ith set of alternative parameters.

(v) We calculate the change in delay duration with respect to the delay duration of the cell cycle model which is shown in Fig. 3.11:

$$\Delta\tau_i = \tau_o - \tau_{ri}, \tag{3.35}$$

where τ_o is the delay duration caused by the cell cycle model (with the set of parameters listed in Table A.2 of Appendix A) in response to 1 M NaCl, and τ_{ri} is the delay duration caused by 1 M NaCl with the ith set of alternative parameters.

(vi) Finally, we normalise each $\Delta\tau_i$ with respect to $|\tau_o|$: $\frac{\Delta\tau_i}{|\tau_o|}$.

The results for each phase are shown in Figs. 3.20a–d. The y-axis shows $\frac{\Delta\tau_i}{|\tau_o|}$ and the x-axis represents i. Therefore, each blue point in Fig. 3.20 depicts the relative delay duration difference, $\frac{\Delta\tau_i}{|\tau_o|}$, for the ith set parameters. Note that each simulation ($4 \times 100 = 400$ in total) is performed for the time span of 300 min. If the cell cycle

Table 3.2 Four different groups of parameters used in sensitivity analysis

Group	Stage of cell cycle	Parameters
I	G1 phase	k_{hpd3c1}; k_{hpdib2}; k_{hpasb2}; k_{hpppc1}; k_{hd3c1}; k_{hdib5}; k_{hasb5}; k_{hdib2}; k_{hasb2}; k_{hppc1}; k_{h1ppc1}; k_{hpkpc1}; kk_{ash}; kkk_{ash1}; kk_{ash1}; $k_{Hog1Cln3}$; $kd_{Hog1Cln3}$; $kd_{Hog1Cln2}$; $kd_{Hog1mass}$; $n_{Hog1Cln3}$; $n_{Hog1Cln2}$; $n_{Hog1Sic1}$
II	S phase	k_{hpd3c1}; k_{hpdib2}; k_{hpasb2}; k_{hpppc1}; k_{hd3c1}; k_{hdib5}; k_{hasb5}; k_{hdib2}; k_{hasb2}; k_{hppc1}; k_{h1ppc1}; k_{hpkpc1}; kk_{ash}; kkk_{ash1}; kk_{ash1}; J_{Hsl1}; k_{hsl1}; J_{iwee}; $kk_{Hsl1Hsl7}$; $kkk_{Hsl1Hsl7}$; $kkd_{Hsl1Hsl7}$; $kd_{Hog1mass}$; $kd_{Hog1Clb5}$
III	DNA re-replication window	k_{hpd3c1}; k_{hpdib2}; k_{hpasb2}; k_{hpppc1}; k_{hd3c1}; k_{hdib5}; k_{hasb5}; k_{hdib2}; k_{hasb2}; k_{hppc1}; k_{h1ppc1}; k_{hpkpc1}; kk_{ash}; kkk_{ash1}; $n_{Hog1Sic1}$; $kd_{Hog1Clb2}$; $kd_{Hog1Clb5}$; k_{pd3c1}; k_{pdib2}; k_{pd3f6}; k_{pdif2}; k_{pasb2}; k_{pasf2}; k_{pppc1}; k_{pppf6}; $n_{Hog1Clb2}$; $kd_{Hog1Clb5}$
IV	M phase	k_{hpd3c1}; k_{hpdib2}; k_{hpasb2}; k_{hpppc1}; k_{hd3c1}; k_{hdib5}; k_{hasb5}; k_{hdib2}; k_{hasb2}; k_{hppc1}; k_{h1ppc1}; k_{hpkpc1}; kk_{ash}; kkk_{ash1}; $n_{Hog1Sic1}$; $kd_{Hog1Clb2}$; $kd_{Hog1Clb5}$; k_{pd3c1}; k_{pdib2}; k_{pd3f6}; k_{pdif2}; k_{pasb2}; k_{pasf2}; k_{pppc1}; k_{pppf6}; $n_{Hog1Clb2}$

model does not divide within the time span of 300 min, we stop the simulation and represent it by a red point in Fig. 3.20, i.e. these cells are permanently arrested.

The results for each of the cell cycle phases show that the delay duration is very robust against variations in the parameters. See Figs. 3.20a (G1 phase), 3.20b (S phase), 3.20c (DNA re-replication window) and 3.20d (M phase). This is especially true for the G1 and S phase: The mean of $\frac{\Delta \tau_i}{|\tau_0|}$ and the respective standard deviation are 0.0636 and 0.503 during the G1 phase. The mean of $\frac{\Delta \tau_i}{|\tau_0|}$ and the standard deviation during the S phase are 0.056 and 0.504 respectively. In the case of the DNA re-replication window, there are a few random sets which yield a non-dividing cell. However the general results show that the predicted delay is robust; with its average value and standard deviation respectively equal to 0.253 and 0.762.

The delay duration is sensitive to the value of parameters related to the M phase. Note that the model predicts an acceleration of the cell cycle, instead of an arrest, if stress is applied during the M phase. 37 % of the chosen random sets show an arrest of the cell cycle instead of an acceleration (red dots in Fig. 3.20d), indicating the sensitivity of the model to the parameters in the M phase. In all those sets, the parameter governing the downregulation of *CLB2* by Hog1PP ($k_{dHog1Clb2}$) happens to be quite small compared to the estimated value. This shows that the prediction of accelerated exit from mitosis is very sensitive to the value of this parameter. In fact, quite small value of $k_{dHog1Clb2}$ causes upregulation of *CLB2* instead of downregulation in the presence of osmotic stress. It has been reported that strong overexpression of *CLB2* arrests the cell cycle in M phase [19]. Therefore the fact that the model prediction is sensitive to this parameter is supported by experimental observation [19].

3.6 Sensitivity Analysis of the Model

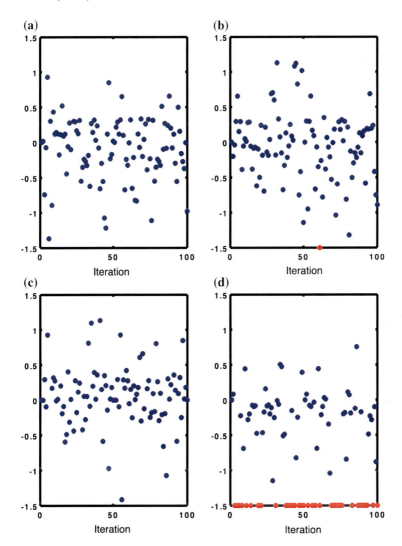

Fig. 3.20 Parameter sensitivity analysis. 100 sets of logarithmically distributed random parameters are generated for each groups of parameters mentioned in Table 3.2 (see text for details of generation of these sets of random parameters). 1 M NaCl is applied at four different stages of the cell cycle; (**a**) G1 phase – at $t = 20$ min, (**b**) S phase – at $t = 50$ min, (**c**) DNA replication window – at $t = 75$ min, (**d**) M phase – at $t = 90$ min. The y-axis shows $\frac{\Delta T_i}{|T_0|}$ and the x-axis represents the iteration index (i). Each blue dot represents $\frac{\Delta T_i}{|T_0|}$ caused by simulation of the model with the ith set of alternative parameters. The red dots represent the cells which are not dividing with the corresponding set of parameters. Their value of $\frac{\Delta T_i}{|T_0|}$ are set to -1.5 for visualisation. See text for further details

In summary the sensitivity analysis shows that the predictions of the model are robust with respect to the variation of the parameters. Previous investigation on the robustness of the cell cycle despite the fluctuations of intracellular parameters confirms our investigation [31].

3.7 Model Validation in Laboratory

To validate the series of predictions of the model, at the moment of writing this work a set of experiments is being carried out by our collaborators. We synchronise the population of cells by the α-factor. We then divide this population of cells into 20 groups. We label each group from 1 to 20 and stress each individual group every 5 min after release from the alpha factor with 1 M of NaCl, i.e. the first group is stressed with 1 M NaCl after 5 min, the second groups is stressed with 1 M NaCl after 10 min, and so on. The last group is stressed with 1 M NaCl after 100 min of being released from the α-factor. We harvest each individual group every 10 min after application of the stress and put it in the flow cytometry device (FACS) to study the progress of the cells through their cell cycle. These experiments will (i) investigate the duration of the delay through the entire cell cycle, (ii) validate the possibility of DNA re-replication without cell division for cells which are at late S phase or early G2/M phase at onset of osmotic stress (in this case a peak at 4N is expected to be observed), and (iii) verify the acceleration in the exit from mitosis.

3.8 Discussion

We have presented a comprehensive model that describes how osmotic stress influences the cell cycle machinery throughout the entire cell cycle [45]. Our model integrates recent experimental findings of the interaction of the osmotic stress response and cell cycle networks across different cell cycle phases. By considering the whole picture, rather than focusing on a specific cell cycle phase, the model is able to unveil mechanisms that emerge as a consequence of the multiple interactions between different parts of the cell cycle and osmotic stress response. The model makes a series of novel predictions and provides mechanisms that explain further experimental findings which lacked explanation so far. Its two main predictions are: (i) upon osmotic stress in late S or early G2/M phase, cells undergo a second incidence of DNA replication before mitosis, (ii) cells stressed at late G2/M phase have an accelerated exit from mitosis and get arrested in the next cell cycle.

3.8.1 The Predictions of the Model are Supported by Various Biological Observations

In untreated cells, DNA re-replication is prevented due to the inhibitory role of Cdc28-Clb5 in the assembly of the pre-replicative complex. When cells are osmotically stressed at the end of the S phase or early G2/M phase, however, the activity of Cdc28-Clb5 is downregulated by Hog1PP. This cancels the inhibition of Cdc28-Clb5 on the pre-replicative complex formation, and thereby causes increased activity of Cdc6 and Cdc14. Then, when Cdc28-Clb5 recovers, after Hog1PP returns to its basal level, a second incidence of DNA re-replication is initiated before mitosis. Note that the level of Cdc6 is lower than in the first incidence of DNA replication, but there is an experimental evidence that low Cdc6 levels are sufficient to licence replication origins and transfer the cell to the S phase [44]. Importantly, the model identifies the mechanisms responsible for DNA re-replication; our results indicate that by blocking the Hog1PP interaction with Sic1, DNA re-replication should be inhibited [45].

The other prediction derived from the model is that Hog1PP may exert another, as yet undetected, level of control on licensing factors to prevent re-replication of DNA. Interestingly, in human cells the licensing protein Cdt1, which assists the assembly of the pre-replicative complex, is phosphorylated by the stress-activated mitogen-activated protein (MAP) kinases. Phosphorylated Cdt1 is then rapidly degraded, thereby inhibiting DNA re-replication upon osmotic stress [11]. In S. cerevisiae Cdt1 is also present, but to our knowledge, it has not been reported that it is phosphorylated by Hog1PP.

The second main prediction of the model, namely the accelerated exit from mitosis, has, to our knowledge, not yet been tested experimentally for S. cerevisiae. However, recent studies on the influence of osmotic stress on dividing leaf cells in the model plant Arabidopsis thaliana also shows early exit from mitosis [16, 53], as the model predicts for budding yeast [45]. If this prediction is not confirmed experimentally in S. cerevisiae, it would strongly indicate that a key component linking the M phase and osmotic stress response is missing; our model suggests that this component should hinder the Clb2 inactivation mechanisms to inhibit accelerated exit from mitosis.

Furthermore, we find that cell cycle progression is delayed approximately linearly with stress dose if the cells are stressed at the G1, S and early G2/M phase. In contrast, cells undergo an accelerated exit from mitosis in a stress dose-independent manner when stressed at late G2/M phase.

3.8.2 The Model Revealed Mechanisms for the Response of the Cell to Osmotic Stress

There is a common mechanism responsible for the delay observed in cells stressed in G1, S and early G2/M phase, even though the details differ from phase to phase. In all cases, interactions of Hog1PP with different cell cycle network components lead to an extended active interval of the transcription factor complexes SBF and MBF. In contrast, the activity of Hog1PP during the G2/M phase causes a reduced active time interval of the transcription factor Mcm1. In the case of cells stressed during the late G2/M phase, it is noteworthy that even though accelerated exit from mitosis is stress dose independent, they are arrested in the next cycle, and the duration of that arrest is indeed dose dependent.

Moreover, the model provides a mechanism that explains why MEN (Mitotic Exit Network) temperature sensitive mutant cells undergo mitosis under osmotic stress [45]. It has been suggested that $cdc15^{ts}$ cells can progress through mitosis due to increased Cdc14 activity mediated by Hog1PP [47]. We find, in accordance with the observation by Grandin et al., that Cdc14 is not the main responsible component for that [27]. This model indicates that the stabilisation of Sic1 by Hog1PP is the key mechanism that drives the mitotic exit in $cdc15^{ts}$ cells under osmotic stress. Therefore, stabilisation of Sic1 by Hog1PP across all cell cycle phases, seems to be the most prominent biochemical event in the interaction between osmotic stress and cell cycle network across the entire cell cycle [45].

3.8.3 The Relevance of the Model Predictions for Other Eukaryotes

Our integrative mathematical model is built based on the molecular mechanisms of the model organism *S. cerevisiae*. However, the molecular basis of the control of two crucial events of the cell cycle, DNA replication and segregation, is highly conserved in eukaryotes. The role of CDKs in the control of the cell cycle is universal [43]. Also HOG MAPK network is highly conserved among eukaryotes. The activity of the HOG MAPK network regulates the activity of CDKs by: (i) altering the transcription of the cyclin partners of CDK (Cln3, Cln2, Clb5 and Clb2), (ii) prolonging the phosphorylation of CDK (mediated by Swe1) and (iii) accumulating the cyclin kinase inhibitor (Sic1). Hence, the cellular predictions of the model are relevant for other eukaryotes.

One of the predictions of the model, namely that osmotic stress causes accelerated exit of mitosis, has been recently experimentally observed by Skirycz et al. in *Arabidopsis thaliana* [53]. This observation validates this postulations and confirms the predictive power of the model.

One should keep in mind that this model has been developed for doses of osmotic stress less than 1 M NaCl. For higher doses of stress, other links and components, not identified yet, may be involved in cellular response to osmotic stress. Moreover, the repair mechanisms for DNA replication errors are not included in the model. Also note that the model is built to unveil the structure of the regulatory mechanisms of the cell in response to osmotic stress rather than to make an exact quantitative prediction

3.8 Discussion

of the levels of the proteins involved. In order to achieve that, further experiments and data fitting is necessary.

In summary, this model provides a series of novel predictions for the interactions between the cell cycle and the osmotic stress response, which suggest new experiments.

References

1. M.A. Adrover, Z. Zi, A. Duch, J. Schaber, A. Gonzalez-Novo, J. Jimenez, M. Nadal-Ribelles, J. Clotet, E. Klipp, F. Posas, Time-dependent quantitative multicomponent control of the G1-S network by the stress-activated protein kinase Hog1 upon osmostress. Sci. Signal. **4**(192), 63 (2011)
2. M.R. Alexander, M. Tyers, M. Perret, B.M. Craig, K.S. Fang, M.C. Gustin, Regulation of cell cycle progression by Swe1p and Hog1p following hypertonic stress. Mol. Biol. Cell **12**(1), 53–62 (2001)
3. A. Amon, M. Tyers, B. Futcher, K. Nasmyth, Mechanisms that help the yeast cell cycle clock tick: G2 cyclins transcriptionally activate G2 cyclins and repress G1 cyclins. Cell **74**(6), 993–1007 (1993)
4. M. Barberis, E. Klipp, M. Vanoni, L. Alberghina, Cell size at S phase initiation: an emergent property of the G1/S network. PLoS Comput. Biol. **3**(4), 649–666 (2007)
5. A.J. Bardin, M.G. Boselli, A. Amon, Mitotic exit regulation through distinct domains within the protein kinase Cdc15. Mol. Cell. Biol. **23**(14), 5018–5030 (2003)
6. R.D. Basco, M.D. Segal, S.I. Reed, Negative regulation of G1 and G2 by S-phase cyclins of Saccharomyces cerevisiae. Mol. Cell. Biol. **15**(9), 5030–5042 (1995)
7. M. Bäumer, G.H. Braus, S. Irniger, Two different modes of cyclin Clb2 proteolysis during mitosis in Saccharomyces cerevisiae. FEBS Lett. **468**(2–3), 142–148 (2000)
8. G. Bellí, E. Garí, M. Aldea, E. Herrero, Osmotic stress causes a G1 cell cycle delay and downregulation of Cln3/Cdc28 activity in Saccharomyces cerevisiae. Mol. Microbiol. **39**(4), 1022–1035 (2001)
9. R.N. Booher, R.J. Deshaies, M.W. Kirschner, Properties of Saccharomyces cerevisiae wee1 and its differential regulation of p34CDC28 in response to G1 and G2 cyclins. EMBO J. **12**(9), 3417–3426 (1993)
10. J.L. Burton, M.J. Solomon, Hsl1p, a Swe1p inhibitor, is degraded via the anaphase-promoting complex. Mol. Cell. Biol. **20**(13), 4614–4625 (2000)
11. S. Chandrasekaran, T.X. Tan, J.R. Hall, J.G. Cook, Stress-stimulated mitogen-activated protein kinases control the stability and activity of the Cdt1 DNA replication licensing factor. Mol. Cell. Biol. **31**(22), 4405–4416 (2011)
12. K.C. Chen, A. Csikasz-Nagy, B. Gyorffy, J. Val, B. Novak, J.J. Tyson, Kinetic analysis of a molecular model of the budding yeast cell cycle. Mol. Cell. Biol. **11**(1), 369–391 (2000)
13. K.C. Chen, L. Calzone, A. Csikasz-nagy, F.R. Cross, B. Novak, J.J. Tyson, Integrative analysis of cell cycle control in budding yeast. Mol. Cell. Biol. **15**(August), 3841–3862 (2004)
14. V.J. Cid, M.J. Shulewitz, K.L. McDonald, J. Thorner, Dynamic localization of the Swe1 regulator Hsl7 during the Saccharomyces cerevisiae cell cycle. Mol. Cell. Biol. **12**(6), 1645–1669 (2001)
15. A. Ciliberto, B. Novak, J.J. Tyson, Mathematical model of the morphogenesis checkpoint in budding yeast. J. Cell Biol. **163**(6), 1243–1254 (2003)
16. H. Claeys, A. Skirycz, K. Maleux, D. Inze, DELLA Signaling mediates stress-induced cell differentiation in Arabidopsis leaves through modulation of APC/C activity. Plant Physiol. **159**(2), 739–747 (2012)

17. J. Clotet, X. Escoté, M.A. Adrover, G. Yaakov, E. Garí, M. Aldea, E. de Nadal, F. Posas, Phosphorylation of Hsl1 by Hog1 leads to a G2 arrest essential for cell survival at high osmolarity. EMBO J. **25**(11), 2338–2346 (2006)
18. F.R. Cross, V. Archambault, M. Miller, M. Klovstad, Testing a mathematical model of the yeast cell cycle. Mol. Biol. Cell **13**(1), 52–70 (2002)
19. F.R. Cross, L. Schroeder, M. Kruse, K.C. Chen, Quantitative characterization of a mitotic cyclin threshold regulating exit from mitosis. Mol. Biol. Cell **16**(5), 2129–2138 (2005)
20. C. Dahmann, J.F. Diffley, K.A. Nasmyth, S-phase-promoting cyclin-dependent kinases prevent re-replication by inhibiting the transition of replication origins to a pre-replicative state. Curr. Biol. **5**(11), 1257–1269 (1995)
21. L. Dirick, T. Böhm, K. Nasmyth, Roles and regulation of Cln-Cdc28 kinases at the start of the cell cycle of Saccharomyces cerevisiae. EMBO J. **14**(19), 4803–4813 (1995)
22. P.R. Dohrmann, W.P. Voth, D.J. Stillman. Role of negative regulation in promoter specificity of the homologous transcriptional activators Ace2p and Swi5p. Mol. Cell. Biol. **16**(4), 1746–1758 (1996)
23. C.B. Epstein, F.R. Cross, Genes that can bypass the CLN requirement for Saccharomyces cerevisiae cell cycle START. Mol. Cell. Biol. **14**(3), 2041–2047 (1994)
24. X. Escoté, M. Zapater, J. Clotet, F. Posas, Hog1 mediates cell-cycle arrest in G1 phase by the dual targeting of Sic1. Nat. Cell Biol. **6**(10), 997–1002 (2004)
25. S. Ghaemmaghami, W.-K. Huh, K. Bower, R.W. Howson, A. Belle, N. Dephoure, E.K. O'Shea, J.S. Weissman, Global analysis of protein expression in yeast. Nature **425**(6959), 737–741 (2003)
26. A. Goldbeter, D.E. Koshland, An amplified sensitivity arising from covalent modification in biological systems. PNAS **78**(11), 6840–6844 (1981)
27. N. Grandin, A. De Almeida, M. Charbonneau, The Cdc14 phosphatase is functionally associated with the Dbf2 protein kinase in Saccharomyces cerevisiae. Mol. Gen. Genet. **258**(1–2), 104–116 (1998)
28. T. Izumi, D.H. Walker, J.L. Maller, Periodic changes in phosphorylation of the Xenopus Cdc25 phosphatase regulate its activity. Mol. Biol. Cell **3**(8), 927–939 (1992)
29. S.L. Jaspersen, J.F. Charles, R.L. Tinker-Kulberg, D.O. Morgan, A late mitotic regulatory network controlling cyclin destruction in Saccharomyces cerevisiae. Mol. Biol. Cell **9**(10), 2803–2817 (1998)
30. S. Jensen, M. Geymonat, L.H. Johnston, Mitotic exit: delaying the end without FEAR. Curr. Biol. **12**(6), R221–223 (2002)
31. K. Kaizu, H. Moriya, H. Kitano, Fragilities caused by dosage imbalance in regulation of the budding yeast cell cycle. PLoS Genet. **6**(4), e1000919 (2010)
32. E. Klipp, B. Nordlander, R. Kröger, P. Gennemark, S. Hohmann, Integrative model of the response of yeast to osmotic shock. Nat. Biotechnol. **23**(8), 975–982 (2005)
33. D.J. Lew, Cell-cycle checkpoints that ensure coordination between nuclear and cytoplasmic events in Saccharomyces cerevisiae. Curr. Opin. Genet. Dev. **10**(1), 47–53 (2000)
34. D.J. Lew, The morphogenesis checkpoint: how yeast cells watch their figures. Curr. Opin. Cell Biol. **15**(6), 648–653 (2003)
35. D. McCollum, K.L. Gould, Timing is everything: regulation of mitotic exit and cytokinesis by the MEN and SIN. Trends Cell Biol. **11**(2), 89–95 (2001)
36. J.N. McMillan, M.S. Longtine, R. Sia, C.L. Theesfeld, E.S. Bardes, J.R. Pringle, D.J. Lew, The morphogenesis checkpoint in Saccharomyces cerevisiae: cell cycle control of Swe1p degradation by Hsl1p and Hsl7p. Mol. Cell. Biol. **19**(10), 6929–6939 (1999)
37. J.N. McMillan, C.L. Theesfeld, J.C. Harrison, E.S.G. Bardes, D.J. Lew, Determinants of Swe1p degradation in Saccharomyces cerevisiae. Mol. Biol. Cell **13**(10), 3560–3575 (2002)
38. J.N. Mcmillan, C.L. Theesfeld, J.C. Harrison, E.S.G. Bardes, D.J. Lew, Determinants of Swe1p degradation in Saccharomyces cerevisiae. Mol. Biol. Cell **13**, 3560–3575 (2002)
39. M.D. Mendenhall, A.E. Hodge, Regulation of Cdc28 cyclin-dependent protein kinase activity during the cell cycle of the yeast Saccharomyces cerevisiae. Microbiol. Mol. Biol. Rev. **62**(4), 1191–1243 (1998)

References

40. D.O. Morgan, Regulation of the APC and the exit from mitosis. Nat. Cell Biol. **1**(2), E47–53 (1999)
41. P. Nash, X. Tang, S. Orlicky, Q. Chen, F.B. Gertler, M.D. Mendenhall, F. Sicheri, T. Pawson, M. Tyers, Multisite phosphorylation of a CDK inhibitor sets a threshold for the onset of DNA replication. Nature **414**(6863), 514–521 (2001)
42. V.Q. Nguyen, C. Co, J.J. Li, Cyclin-dependent kinases prevent DNA re-replication through multiple mechanisms. Nature **411**(6841), 1068–1073 (2001)
43. P. Nurse, Regulation of the eukaryotic cell cycle. Eur. J. Cancer A **33**(7), 1002–1004 (1997)
44. S. Piatti, T. Bohm, J.H. Cocker, J.F.X. Diffley, K. Nasmyth, Activation of S-phase-promoting CDKs in late G1 defines a 'point of no return' after which Cdc6 synthesis cannot promote DNA replication in yeast. Genes Dev. **10**(12), 1516–1531 (1996)
45. E. Radmaneshfar, D. Kaloriti, M.C. Gustin, N.A.R Gow, A.J.P Brown, C. Grebogi, M.C. Romano, M. Thiel, From START to FINISH: the influence of osmotic stress on the cell cycle. PLoS ONE **8**(7), e68067 (2013)
46. V. Reiser, H. Ruis, G. Ammerer, Kinase activity-dependent nuclear export opposes stress-induced nuclear accumulation and retention of Hog1 Mitogen-activated Protein Kinase in the budding yeast Saccharomyces cerevisiae. Mol. Biol. Cell **10**(4), 1147–1161 (1999)
47. V. Reiser, K.E. D'Aquino, E. Ly-Sha, A. Amon, The stress-activated mitogen-activated protein kinase signaling cascade promotes exit from mitosis. Mol. Biol. Cell **17**(7), 3136–3146 (2006)
48. E. Schwob, T. Böhm, M.D. Mendenhall, K. Nasmyth, The B-type cyclin kinase inhibitor p40SIC1 controls the G1 to S transition in S. cerevisiae. Cell **79**(2), 233–244 (1994)
49. M. Shirayama, Y. Matsui, A. Toh-e, Dominant mutant alleles of yeast protein kinase gene CDC15 suppress the Lte1 defect in termination of M phase and genetically interact with CDC14. Mol. Gen. Genet. **251**(2), 176–185 (1996)
50. R. Sia, H. Herald, D.J. Lew, Cdc28 tyrosine phosphorylation and the morphogenesis checkpoint in budding yeast. Mol. Biol. Cell **7**(11), 1657–1666 (1996)
51. R. Sia, E.S. Bardes, D.J. Lew, Control of Swe1p degradation by the morphogenesis checkpoint. EMBO J. **17**(22), 6678–6688 (1998)
52. K.J. Simpson-Lavy, J. Sajman, D. Zenvirth, M. Brandeis, APC/C Cdh1 specific degradation of Hsl1 and Clb2 is required for proper stress responses of S. cerevisiae. Cell Cycle **8**(18), 3006–3012 (2009)
53. A. Skirycz, H. Claeys, S. de Bodt, A. Oikawa, S. Shinoda, M. Andriankaja, K. Maleux, N.B. Eloy, F. Coppens, S. Yoo, K. Saito, D. Inze, Pause-and-stop: the effects of osmotic stress on cell proliferation during early leaf development in Arabidopsis and a role for ethylene signaling in cell cycle arrest. Plant Cell **23**(5), 1876–1888 (2011)
54. F. Stegmeier, R. Visintin, A. Amon, Separase, polo kinase, the kinetochore protein Slk19, and Spo12 function in a network that controls Cdc14 localization during early anaphase. Cell **108**(2), 207–220 (2002)
55. F. Stegmeier, A. Amon, Closing mitosis: the functions of the Cdc14 phosphatase and its regulation. Annu. Rev. Genet. **38**, 203–232 (2004)
56. T.T. Su, P.J. Follette, P.H. O'Farrell, Qualifying for the license to replicate. Cell **81**(6), 825–828 (1995)
57. U. Surana, H. Robitsch, C. Price, T. Schuster, I. Fitch, B. Futcher, K. Nasmyth, The role of CDC28 and cyclins during mitosis in the budding yeast S. cerevisiae. Cell **65**(1), 145–61 (1991)
58. C.L. Theesfeld, T.R. Zyla, E.G.S. Bardes, D.J. Lew, A monitor for bud emergence in the yeast morphogenesis checkpoint. Mol. Biol. Cell **14**, 3280–3291 (2003)
59. J.H. Toyn, A.L. Johnson, J.D. Donovan, W.M. Toone, L.H. Johnston, The Swi5 transcription factor of Saccharomyces cerevisiae has a role in exit from mitosis through induction of the cdk-inhibitor Sic1 in telophase. Genetics **145**(1), 85–96 (1997)
60. E.E. Traverso, C. Baskerville, Y. Liu, W. Shou, P. James, R.J. Deshaies, H. Charbonneau, Characterization of the Net1 Cell Cycle-dependent Regulator of the Cdc14 Phosphatase from Budding Yeast. J. Biol. Chem. **276**(24), 21924–21931 (2001)
61. M. Tyers, G. Tokiwa, B. Futcher, Comparison of the Saccharomyces cerevisiae G1 cyclins: Cln3 may be an upstream activator of Cln1, Cln2 and other cyclins. EMBO J. **12**(5), 1955–1968 (1993)

62. R. Visintin, S. Prinz, A. Amon, CDC20 and CDH1: a family of substrate-specific activators of APC-dependent proteolysis. Science **278**(5337), 460–3 (1997)
63. G. Yaakov, A. Duch, M. Garcí-Rubio, J. Clotet, J. Jimenez, A. Aguilera, F. Posas, The stress-activated protein kinase Hog1 mediates S phase delay in response to osmostress. Mol. Biol. Cell **20**(15), 3572–3582 (2009)
64. W. Zachariae, K. Nasmyth, Whose end is destruction: cell division and the anaphase-promoting complex. Genes Dev. **13**(16), 2039–2058 (1999)
65. Z. Zi, W. Liebermeister, E. Klipp, A quantitative study of the Hog1 MAPK Response to fluctuating osmotic stress in Saccharomyces cerevisiae. PLoS ONE **5**(3), e9522 (2010)

Chapter 4
Boolean Model of the Cell Cycle Response to Stress

4.1 Introduction

Understanding complex biological systems, e.g. the cell cycle, requires not only sophisticated experimental techniques but also adequate mathematical models. Many different mathematical approaches, from quantitive to qualitative, from continuous to discrete, have been applied to study the cell in different environmental conditions. In this chapter, we introduce a second complementary modelling approach to study the response of the cell cycle to osmotic and alpha-factor signal: we construct a Boolean network which describes the dynamical behaviour of the cell cycle response to multiple extracellular signals.

Diverse methods for modelling cellular regulatory networks such as ordinary differential equations (ODE), stochastic differential equations, Petri nets and Boolean networks have been put forward [2, 4, 6, 9, 10, 19]. The choice of a suitable method depends on the aspects of the system under investigation, the type of available data and also on the question one wants to address. Boolean network modelling can be thought of as the coarse-grained limit of more detailed approaches, such as differential equations, to study systems with complicated dynamics [5].

The ODE modelling approach used in Chap. 3, captures the underlying reaction kinetics in terms of rates, and concentrations and contains the time evolution details of the system. Often these types of models are highly predictive and are a complementing alternative mathematical workbench to experiments for further investigation of biological systems in great details [4, 8, 12]. These models, however, incorporate a large number of biochemical parameters – which in turn requires many experiments to be conducted [3, 4, 8].

Alternatively, one can construct more abstract models in which the regulatory networks are simplified either in terms of their dynamics or the topology of their interaction network. Different approaches can be adopted to achieve a simple model which still captures the behaviour of the underlying system correctly. One of them is to build a minimal ODE model to explain the experimentally observed behaviour of

E. Radmaneshfar, *Mathematical Modelling of the Cell Cycle Stress Response*,
Springer Theses, DOI: 10.1007/978-3-319-00744-1_4,
© Springer International Publishing Switzerland 2014

the cell cycle. The number of components and interactions is reduced considerably, and often the main features of the system can still be described correctly [18].

An alternative simplification is to use a class of discrete dynamical systems, or a Boolean network. A Boolean network consists of a set of binary-state nodes whose values are determined by the topology of the network and the states of their neighbouring nodes (1 or 0) [7]. The first Boolean network to study gene regulatory networks was introduced by Kauffman [7]. In these networks nodes represented genes/proteins and edges describe the biological interactions between them. The state of a node is *one* when it is active, and *zero* when it is inactive. Recent applications of this framework to model different biological systems such as control of the cell cycle [6, 9], T-cell receptor signalling [15], and the segment polarity network of the fruit fly [1], demonstrate how Boolean networks serve as useful *in silico* tools in different cellular regulatory networks. Similar to the ODE modelling framework, the starting point to construct a Boolean network is to establish the wiring diagram of interactions. However, unlike in the ODE modelling approach, no kinetics details are needed. The Boolean networks reveal the dynamical properties and the sequence of events of the regulatory networks.

In this chapter we develop a Boolean model to explore the dynamical behaviour of budding yeast in response to osmotic and alpha-factor signal. Our model predicts that the states of the cell cycle trajectory are attracted to either of four fixed points under osmotic stress, i.e. the cell cycle will be arrested in one of four distinct states. This model also describes the sequence of states towards the fixed point, and the activity of all cell cycle components at this point. It also shows that the state of the cell at the onset of stress dictates the arrest state of the cell.

The model reveals that the cells which are arrested in either of these four fixed points return to the cell cycle trajectory after removal of the osmotic stress. Therefore, the cell can recover from these osmotic stresses. Furthermore, our model illustrates the state transitions during the recovery from the stress.

We also show that the cell cycle trajectory lies in the largest basin of the state transition network, i.e. many states converge towards it. We hypothesise that this is the result of an evolutionary process. The calculations also show that one of the four osmotic stress fixed points has a relatively small basin. This basin, however, can be reached by suitable experimental conditions and play a major role in the stress response. Hence, rather than the mere size of the basin, it is crucial to consider whether typical environmental/experimental fluctuations can take the cell to the states in a particular basin.

One key result of the model regards a standard experimental procedure. The response to stresses is often studied in populations of cells, whose cell cycles have been synchronised. One of the most frequently used methods for cell cycle synchronisation is via the α-factor, which arrests the cells in G1 phase. After release from the α-factor the cells go through the next few cycles synchronously. To study the influence of synchronisation on the dynamics of a population of cells which is exposed to osmotic stress, we expand the Boolean model by adding the corresponding links and nodes. The model demonstrates that one of the osmotic stress arrest points cannot be studied via α-factor synchronisation.

4.1 Introduction

Our model also predicts that osmotic stress can *connect* certain nodes of the state transition network, which are disconnected in unstressed conditions, to the cell cycle basin. We call these disconnected nodes *frozen* states. They describe cells in particular states, which are not able to divide in a *normal* environment, i.e. even if the stress is removed. These cells cannot form colonies and are often considered to be *dead*. The application of an osmotic stress reconnects these states to basin of the cell cycle trajectory. These cells can form colonies again and are perfectly healthy. Hence, if we accept the definition that cells that cannot form colonies in unstressed conditions are dead, this leads to the curious consequence that osmotic stress can *revive* cells.

4.2 A Discrete Dynamical Model

In this section, we describe the Boolean model of the cell cycle response to osmotic stress and the α-factor presented in Radmaneshfar et al. [11]. To unveil the dynamical properties of the cell cycle stress response, we build a Boolean network in which each vertex l represents one of the proteins assuming a binary value $s_l(t) \in \{0, 1\}$. The value of $s_l(t)$ depends on the activity status of component l, which is either "on" or "off". Each edge stands for an interaction (activation or inactivation) between the corresponding vertices. For example Cln3 (node 1 in Fig. 4.1) interacts with MBF (node 2 in Fig. 4.1) and SBF (node 3 in Fig. 4.1). Since these are activation interactions, Cln3 is linked by two solid lines to MBF and SBF (see Fig. 4.1). Note that the activation link (solid line) has weight 1 whereas the inactivation link (dashed line) has weight -1 in Fig. 4.1.

In order to build a Boolean model of *S. cerevisiae*'s cell cycle response to osmotic stress, we start from a recently developed Boolean model of the cell cycle by Li et al. [9], which contains 11 cell cycle core components [16]. As a second step, we add three further cell cycle proteins (Swe1, Mih1 and Hsl1), which will later allow us to link the stress response into the Boolean cell cycle model. To this network of 14 components we add another node, Hog1PP, which is the downstream protein kinase of the osmotic stress signalling pathway and interacts with Cln2, Cln3, Sic1, Clb5, Clb2 and Hsl1, which are cell cycle regulated (see Chap. 2 for details of interactions of Hog1PP with the cell cycle components). Finally, we introduce Far1 and its link to Cln2. This allows us to model the response of the cell to the α-factor. Fig. 4.1 illustrates the resulting Boolean network.

The incidence matrix of the combined cell-cycle-osmotic-stress-pheromone model is a 16×16 matrix, $B = [b_{lk}]_{16 \times 16}, l, k \in \{1, 2, \cdots, 16\}$. The elements b_{lk} are called weight factors, and are defined such as, $b_{lk} = 1$ if protein k activates protein l, $b_{lk} = -1$ if protein k inhibits protein l, and finally $b_{lk} = 0$ if there is no link from the protein k to l. For example, since Cln3 activates MBF and SBF, $b_{21} = 1$ and $b_{31} = 1$. Note that in general, $b_{lk} \neq b_{kl}$. The incidence matrix and the connectivity graph (see Fig. 4.1) both present the same information. The incidence matrix of the Boolean model, B, is presented below:

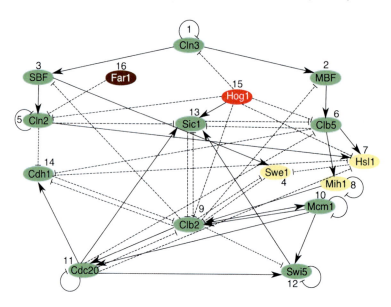

Fig. 4.1 Boolean model of the regulatory network of the cell cycle under two different extracellular signals: (i) osmotic stress and (ii) the α-factor. The Boolean network consists of 16 nodes, 14 of which are cell cycle regulated (they are shown in *green* and *yellow*). The other two nodes, Hog1PP and Far1 (they are shown in *red* and *brown*), do not have any input from the rest of the network. *Solid lines* represent positive regulations and *dashed lines* represent deactivation (inhibition, repression, or degradation)

	Cln3	MBF	SBF	Swe1	Cln2	Clb5	Hsl1	Mih1	Clb2	Mcm1	Cdc20	Swi5	Sic1	Cdh1	Hog1	Far1
Cln3	0	0	0	0	0	0	0	0	0	0	0	0	0	0	−1	0
MBF	1	0	0	0	0	0	0	0	−1	0	0	0	0	0	0	0
SBF	1	0	0	0	0	0	0	0	−1	0	0	0	0	0	0	0
Swe1	0	1	0	0	0	0	0	0	0	0	0	0	0	0	−1	−1
Cln2	0	0	0	0	−1	−1	0	0	−1	0	1	0	0	0	0	0
Clb5	0	0	0	0	0	0	0	0	−1	1	1	0	0	0	0	0
Hsl1	0	0	0	0	0	0	0	0	1	1	0	0	0	0	0	0
Mih1	0	0	1	0	0	0	0	0	0	0	−1	0	−1	0	−1	0
Clb2	0	0	0	0	−1	−1	0	0	−1	0	1	1	0	0	1	0
Mcm1	0	0	0	−1	0	1	0	1	0	1	−1	0	−1	−1	−1	0
Cdc20	0	0	0	0	0	0	0	0	1	0	0	0	0	0	0	0
Swi5	0	1	0	0	0	0	−1	0	−1	0	0	0	0	0	0	0
Sic1	0	0	0	0	1	1	0	0	0	0	−1	0	0	−1	−1	0
Cdh1	0	0	0	0	0	1	0	0	0	0	0	0	0	0	0	0
Hog1	0	0	0	0	0	0	0	0	0	0	0	0	0	0	0	0
Far1	0	0	0	0	0	0	0	0	0	0	0	0	0	0	0	0

The time evolution of the cell cycle components $l \in \{1, \ldots, 14\}$ is described by an iterative process in which the subsequent state of each node is determined by the following Boolean rule:

$$s_l(t+1) = \begin{cases} 1 & \text{if } \sum_k b_{lk} s_k(t) > 0 \\ 0 & \text{if } \sum_k b_{lk} s_k(t) < 0 \\ s_l(t) & \text{if } \sum_k b_{lk} s_k(t) = 0. \end{cases} \quad (4.1)$$

4.2 A Discrete Dynamical Model

The sums calculate the weighted average of all activating and inhibiting inputs to a node. If the overall input is positive the node becomes active, otherwise it becomes inactive. The state of a cell cycle node at time $t + 1$ is hence given by the weighted average of the input nodes at time t. This definition is the same as the one used by Li et al. [9].

Note that Hog1PP and Far1 (nodes 15 and 16 in Fig. 4.1) do not have any incoming links from the cell cycle, since to our knowledge no cell cycle regulation mechanisms for these two components have been reported. Hence, in the model they are independent variables that become active at the onset of the corresponding extracellular signals, and are constant in time. Therefore, the Boolean function for Hog1PP and Far1 is independent of the other components (Fig. 4.1), i.e. $s_l(t + 1) = s_l(t)$ if $l \in \{15, 16\}$.

In every iteration the states of all other nodes of the Boolean network are updated synchronously (for definition of synchronous update see Appendix B) due to intrinsic biological properties of the cell cycle [9]. Note that the set of fixed points, which is central to our subsequent arguments, is independent of the update rule, i.e. synchronous update, asynchronous update (for definition of asynchronous update see Appendix B) and a combination of both all lead to the same set of fixed points (see Appendix B) [14].

This Boolean model reveals the dynamical properties of the cell cycle both in untreated and treated with osmotic stress and the α-factor conditions. It is noteworthy that the dynamical behaviour of this coarse-grained model is consistent with biological observations, as we will show later [6, 9, 11].

Next, we will explain how to construct the state transition matrix of the model. This matrix describes how the overall state of the Boolean network transits to a subsequent state. Then, we will discuss the dynamical response of the cell cycle to osmotic stresses in different experimental setups.

4.3 State Transition Space of the Cell Cycle

In this section we introduce the state transition matrix of the developed Boolean model [11]. It describes transitions between the global states of the Boolean network. This transition matrix can be considered as the adjacency matrix of the state transition graph. Note that this network is much larger than the Boolean one. In fact, every node of the transition network corresponds to the state of the entire Boolean network of N nodes, i.e. each of the 2^N states is represented by a node in the transition network. Hence, the state transition matrix $(T = [t_{ij}])$ is a $2^{16} \times 2^{16}$ square matrix, where $N = 16$ is the number of nodes in the corresponding Boolean network.

We reduce the size of the state transition matrix of the Boolean model by considering the fact that Hog1PP and Far1 are not cell cycle regulated and also order the nodes as shown in Fig. 4.1. Since there is no input link to these two stress nodes, their states remain constant (equal to initial state) as time evolves. As a consequence the $2^{16} \times 2^{16}$ state transition matrix has four $2^{14} \times 2^{14}$ sub-square matrices. Each of which describes one of the four stress combinations, where two stress nodes can be either active or inactive. The first matrix corresponds to the case when both Hog1PP and Far1 are inactive; we call this matrix U (<u>u</u>nstressed). When Hog1PP is active and Far1 is inactive we denote the corresponding matrix by O (<u>o</u>smotic). The third matrix A (<u>a</u>lpha factor) represents when Hog1PP is inactive and Far1 is active. Finally, the fourth matrix C (<u>c</u>ombinatorial) describes the situation when both Hog1PP and Far1 are active. So each of the U, O, A and C matrices models the state transition of the cell cycle in different environmental conditions. Consequently, the state transition matrix T is of the form:

$$T = \begin{bmatrix} U & 0 & 0 & 0 \\ 0 & O & 0 & 0 \\ 0 & 0 & A & 0 \\ 0 & 0 & 0 & C \end{bmatrix}_{2^{16} \times 2^{16}}. \quad (4.2)$$

Next, we need to derive the different state transition matrices (U, O, A and C), which will enable us to study the dynamics of cell cycle in different environmental conditions. The following discussion is exemplified for the state transition matrix U (unstressed), but it is valid for the other blocks as well.

4.3.1 Deriving the State Transition Matrix

The transition matrix describes the time evolution of the Boolean network. If a transition from state j to state i is possible by iteration with the Boolean function (Eq. 4.1), the element u_{ij} of U is 1; otherwise it is 0.

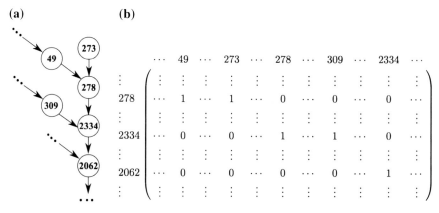

Fig. 4.2 (a) An example of a state transition network. Numbers inside each *circle* are the decimal representation of the corresponding states of Boolean network nodes. (b) State transition matrix of the graph shown in part a. See text for details of the calculation

4.3 State Transition Space of the Cell Cycle 77

It is convenient to map the binary representation of each state of the Boolean network $(s_1\ s_2\ \cdots\ s_{14})$ to the respective decimal representation, i.e.

$$(s_1\ s_2\ \cdots\ s_{14})_2 = (i)_{10}, \qquad (4.3)$$

where $i \in \{1, \ldots, 2^{14}\}$ and s_l, $l \in \{1, \ldots, 14\}$. We use this decimal representation to label the respective states in the state transition network. Figure 4.2a shows part of a state transition graph and Fig. 4.2b depicts the corresponding state transition matrix. For example, according to Fig. 4.2a state 273 and 49 transit to state 278 as time evolves, therefore, $u_{278,49} = 1$ and $u_{278,273} = 1$.

Following the same argument, we can find the other state transition matrices (O (osmotic), A (α-factor) and C (combinatorial)) and their corresponding networks. The corresponding state transition graphs of the matrices U and O are shown in Figs. 4.4 and 4.5 respectively.

4.3.2 Dynamical Properties of the Cell Cycle State Transition Matrix

The matrix U (unstressed) with elements $u_{i,j}$ $i, j \in \{1, 2, \cdots, 2^{14}\}$ is the adjacency matrix of the state transition graph when there is no stress. The state transition graph of the Boolean model is, in fact, a directed tree. We will discuss the reasons behind this structure later. Since every state transfers to a unique state, $d_j^{out} = \sum_i u_{ij} = 1$. The number of the states being mapped to state i after one time application of Boolean function is $d_i^{in} = \sum_j u_{ij}$. Having obtained the state transition matrix, we are able to study the dynamics of transitions among all states. The set of all possible states $\mathbf{X}$ is finite. Hence, starting from any element of $\mathbf{X}$ the state of the system will necessary reach either a fixed point or a limit cycle after a sufficient number of iterations with U (see Appendix B) [14]. As this state transition graph has the structure of a directed tree, it cannot, by definition, contain any cycles.

As mentioned above, the state transition matrix U (unstressed) captures the time evolution of its corresponding Boolean network. The state $\mathbf{x}$ is represented by a 2^{14} dimensional vector with exactly one non-vanishing component, which is equal to one:

$$\mathbf{x} = \begin{bmatrix} 0 \\ \vdots \\ 1_j \\ \vdots \\ 0 \end{bmatrix} \quad j \in \{1, \ldots, 2^{14}\}. \qquad (4.4)$$

In fact, the state of the Boolean network after the kth iterations starting from any initial condition $\mathbf{x^0}$ can be evaluated by $\mathbf{x^k} = U^k \mathbf{x^0}$. Hence, for the fixed point (**fp**), independently of the value of k, we always have $\mathbf{fp^k} = \mathbf{fp^0} = U^k\mathbf{fp^0}$. Note, that this fixed point equation is an eigenvalue equation to a unit eigenvalue. Therefore, the number of fixed points (n_{fp}) of a Boolean network equals the number of unit eigenvalues of the state transition matrix U (unstressed).

4.3.3 Basin of Attraction

The fixed points allow us to partition the space of all possible states $\mathbf{X}$ to *equivalence* classes which are defined as follows:

Definition Let $\mathbf{X}$ be the set of all states. If $\mathbf{x}$ and $\mathbf{y}$ are in $\mathbf{X}$, $\mathbf{y}$ is a *descendant* of $\mathbf{x}$ if there exists an integer $m \geq 0$ such that $\mathbf{y} = U^m\mathbf{x}$. Two states $\mathbf{w}$ and $\mathbf{z}$ in $\mathbf{X}$ are said to be *equivalent* if they have a common *descendant*.

A fixed point is *descendant* of all states which are being attracted towards it.

Definition The *equivalence class* of each fixed point is called *basin* of that fixed point or *basin of attraction*.

Moreover, to partition $\mathbf{X}$ into its basins it is convenient to first determine the minimal k such that $U^{k+1} = U^k$. The number of the basins of attraction is equal to the number of unit eigenvalues which is equal to the number of fixed points. For an arbitrary initial state, k is greater or equal to the number of iterations after which it reaches its *descendant* fixed point. Note that k is the largest diameter of all basins. Hence, since the state transition graph of U (unstressed) is a directed tree the matrix U^k has n_{fp} nonzero rows. The nonzero elements of each of these n_{fp} nonzero rows constitute their corresponding basin.

4.4 Results

4.4.1 Cell Cycle State Transition

To study the time evolution of the cell cycle we set the initial conditions of the nodes in the Boolean network (see Fig. 4.1) equal to a biological state, which is called START. We multiply the corresponding vector of this state (Eq. 4.4) by the matrix U, until we reach a fixed point. The output of these iterations gives us the sequence of state transition in the cell cycle. Figure 4.3a shows the sequential activation of the cell cycle components according to our Boolean model. The cell cycle components are shown with different colours in Fig. 4.3a for visualisation purposes, and they are active during their corresponding sectors.

As mentioned before, the state transition graph of the cell cycle Boolean model is a directed tree [9, 11]. Figure 4.3b shows the binary representation of the cell cycle

4.4 Results

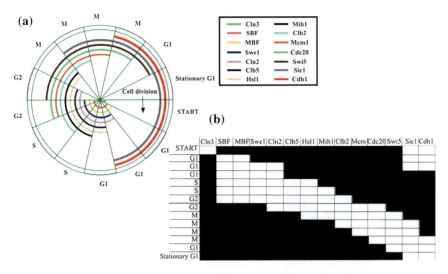

Fig. 4.3 (a) Sequential activity of the cell cycle components during the state transition from START to **fp**$_U$ (stationary G1). The components are shown with different *colours* and they are active during their corresponding sectors. (b) Boolean activity profile of the cell cycle component. Note that Cln3 activity is different in the stationary G1 and START states. Cln3 triggers a cascade of state transitions from START to **fp**$_U$. We did not consider the signal which reverses stationary G1 to START. Hence, we consider one period of the cell cycle. The cascade of events ends in the fixed point **fp**$_U$. Activity of Cln3 returns the cell from that fixed points to START. An active component is shown by *white*, whereas *black* block depicts an inactive component

state transition network. A component is active, 1, when it is shown by white block, and it is inactive, 0, when it is depicted by black. In Fig. 4.3b we have labelled the consecutive states of the cell cycle from START to stationary G1.

If the initial conditions of the Boolean network nodes are the same as the START state, time evolution of the Boolean network brings that state to the end of the cell cycle trajectory (stationary G1 state) and the cell cycle is completed. Note that the Hamming distance[1], i.e. the number of different entries in the binary state vectors, between the first and the last states in the Fig. 4.3 is one. If the cell is in the final state, stationary G1, the onset of the cell division signal will return the cell back to the initial state. This signal activates the Cln3 node, and changes its Boolean value from 0 to 1. Cln3 is the cyclin that triggers cell cycle progression [17]. The exact mechanisms that activate Cln3 are still unknown. Since we are rather interested in the state transitions in one cycle, the cell division signal is not considered explicitly in our model.

---

[1] The Hamming distance between two strings of equal length is the number of positions at which the corresponding symbols are different.

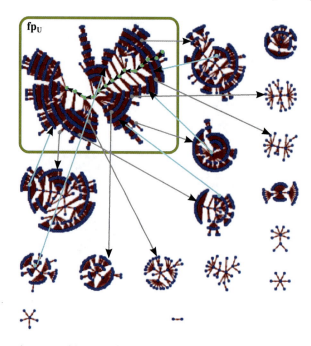

Fig. 4.4 Directed state transition tree of the global set of initial conditions. The cell cycle trajectory is in the largest basin of the attraction (**U**-basin). The nodes of this trajectory are shown in *green* and the edges in *black*. We call the states in the other basins frozen states. Presence of the osmotic stress can bring some of the frozen states back to the U-basin. Examples of these states are shown in *light blue arrows*. Moreover, osmotic stress can kick some states of the U-basin out to the other basins. Hence, they are arrested forever after removal of the stress. Some of these states are shown in *grey arrows*. Some of these states are artificial; they are obtained by mathematically generating all possible combinations of "on /off" states of all constituents (see Sect. 4.4.5 for further details). This graph has been plotted using Mathematica

4.4.2 Osmotic Stress Drives the Cell into One of the Four Fixed Points

The state transition graph of the Boolean network of budding yeast is shown in Fig. 4.4. The state transition matrices, described in the Sect. 4.3, contain all information about the dynamics of the system. Their fixed points, i.e. eigenvectors to a unit eigenvalue, are of particular interest, as they correspond to an *arrested* state, i.e. a state that the cell cannot leave without an external signal. We calculate the eigenvectors to the 17 unit eigenvalues of the matrix U (unstressed) of the Boolean network. These fixed points partition the state space into 17 different basins, i.e. equivalence classes. We denote the basin which contains the cell cycle trajectory **U**-basin to emphasise that it is the basin of the unstressed case. The fixed point of the **U**-basin is denoted by **fp$_U$**.

4.4 Results

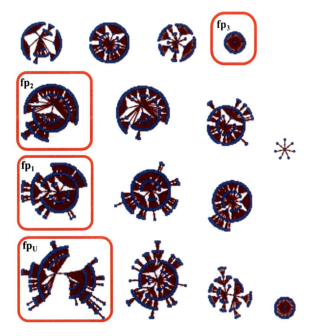

Fig. 4.5 Directed state transition tree of the global set of initial conditions, when the Hog1PP node is active, corresponding graph of matrix O. The states of the cell cycle trajectory converge to either of the basins which are highlighted by a *red box*. This graph has been plotted using Mathematica

Osmotic stress causes activation of the MAPK HOG network, which as a result, activates the Hog1PP node. The activity of the Hog1PP node in turn changes the state transition matrix from U (unstressed) to O (osmotic). The matrix O has 15 unit eigenvalues, and consequently 15 basins (Fig. 4.5). A short calculation shows that states from the unstressed cell cycle trajectory (Fig. 4.3) are now elements of four different basins of O (out of 15).

It is noteworthy that one of these fixed points is fp_U. Osmotic stress halts cells which are stressed upstream of the state *CC7* in Fig. 4.6. Accordingly, the cell is arrested in either of the three fixed points (fp_1, fp_2, fp_3 in Fig. 4.6b). The cells which have passed the state *CC7*, will go through the same state transitions as unstressed cells (compare Fig. 4.6a with 4.6b), until they reach fp_U. Hence, the state of the cells at the onset of the stress dictates their time evolution and ultimately which fixed point they are attracted to.

Moreover, our model also predicts the state transitions from the four fixed points of O (osmotic) to the cell cycle trajectory when the stress is removed. According to the model, a perturbation from the cell cycle trajectory by the osmotic stress can be compensated when the stress node is off (green edges in Fig. 4.6b). The cell cycle trajectory and its response to the osmotic perturbation are depicted in Figs. 4.6a and b. The black edges show the state transitions for the unperturbed case, whereas the red edges illustrate the network's response to the activity of Hog1PP. Finally, the green edges represent the state transitions from the fixed points of O to the cell cycle trajectory, under osmotic condition. Remarkably, the presence of osmotic stress cannot take the cell out of its basin, and all of these perturbations from the limit cycle of the cell cycle can be compensated after the removal of the osmotic stress.

4.4.3 Biological Relevance of the Size of Basins

In Sect. 4.4.2 we have calculated the fixed points of the state transition matrices and discussed their relevance for the interpretation of the cells' stress response. In this section we will study the importance of the respective basins and particularly their size.

We start the discussion with the observation that one of the basins of U (unstressed) is much larger than the others. This is consistent with what has been observed previously in the Boolean model of the cell cycle [9]. Like before, the cell cycle trajectory (see Fig. 4.3 for the cell cycle trajectory) lies in this largest basin (U-basin in Fig. 4.4) [9].

To interpret this result, one should keep in mind that many of the states in the U-basin are to some extent *artificial*; they are obtained by mathematically generating all possible combinations of states of all constituents. How many of them are viable or can be reached by careful experimentation or environmental fluctuation is a challenging question. The special relevance of the U-basin is that the cell cycle trajectory lies in this basin, rather than its mere size. It can be speculated however, that those states that can be reached by typical environmental fluctuations should be preferentially linked to the U-basin on evolutionary time scales, to allow the cells to return to cell cycle trajectory and reproduce.

The size of one of the four cell cycle basins of O (osmotic) (node **fp₃** in Figs. 4.6b and 4.5) is very small compared to the others. But since it is known how to reach that particular basin experimentally, it is a crucial basin for the following discussion.

So rather than the mere size of basin, it is important to know how to reach the states in that basin and whether they correspond to viable cells. We therefore suggest to consider only those states which are *observable*, i.e. can be reached by environmental or experimental perturbations.

4.4.4 Influence of the α-Factor Synchronisation on the Cell Cycle Dynamics

In this section we will discuss an experimental procedure that is frequently used to synchronise populations of cells. In a culture of cells the individuals will be at different stages of their cell cycle and this adds an additional complication to many experiments. Synchronising the cell cycle in a cell population allows us to look at

4.4 Results

more specific settings and still preserves the advantage of large ensembles of cells, which is required for many experimental techniques. This is why many traditional techniques requires to synchronise the population to study the response of cells to environmental stresses at a particular phase. Yet, it is important to understand the influence of the chosen synchronisation technique on the cell cycle dynamics and the state transitions. Ignoring the influence of the synchronisation method on the dynamics may mislead the interpretation of the results.

One of the standard methods of synchronisation of the cell cycle of a culture of budding yeast is via the α-factor. This activates a cascade of events which at the end inhibits Cln2 and arrests the cells in G1 phase.

We first add Far1 node and its corresponding edge to our Boolean model of the cell cycle (see Fig. 4.1). We then investigate the state transitions of the cell cycle when the α-factor is applied by deriving matrix A. The states of the cell cycle trajectory evolve into either of two fixed points which we denote by $\mathbf{fp_1}$ and $\mathbf{fp_U}$ respectively in Fig. 4.6c. Cells in states which are attracted to $\mathbf{fp_U}$ will divide exactly once upon presence of the cell cycle division signal despite the presence of the α-factor. But those which are in $\mathbf{fp_1}$'s basin will get arrested at that fixed point when Far1 is active. The state transitions of the cell during and after release from the α-factor is depicted in Fig. 4.6c.

Interestingly, these two fixed points are also osmotic stress fixed points (Fig. 4.6). Considering the fact that the cell will get arrested in $\mathbf{fp_1}$ when the α-factor is used, we observe that synchronisation with the α-factor will limit the study to those cell cycle states which are downstream of *CC3* in Fig. 4.6. This indicates that using the α-factor for population synchronisation allow studying all osmotic stress fixed points apart from $\mathbf{fp_1}$ and its basin. Therefore, to study $\mathbf{fp_1}$ other methods of population synchronisation must be used.

4.4.5 Osmotic Stress Can Retrieve Some Frozen States to the Cell Cycle Trajectory

The extracellular signals we have considered in this chapter so far, are all *reversible*, in the sense that, once the signal is removed, the cells will resume their cell cycle. That means that the system stays within the U-basin (cell cycle basin). As discussed before, there are 16 other basins which are disconnected from the U-basin (Fig. 4.4). If a stress moves the system into one of those basins, the cells will not recover after the removal of the stress. They will not resume their cell cycles and they will not form colonies. Following a common definition in experimental biology these cells can be considered to be *dead* in spite of their remaining metabolic activity. Our model suggests that some of these frozen states, might be returned to the U-basin if an appropriate experimental procedure e.g., stress, is applied.

Within the framework of our model, the Boolean state of the stress node causes a transition of some frozen states, that are originally outside of the U-basin, into the

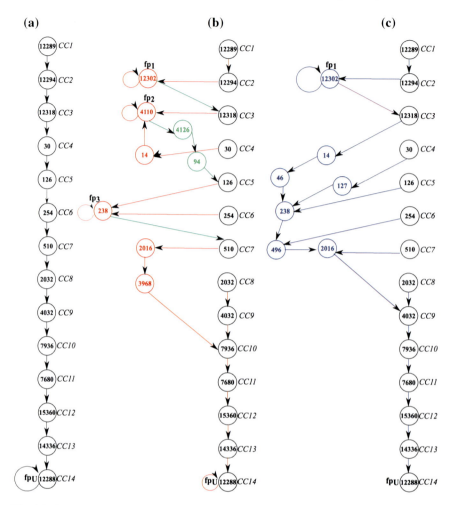

Fig. 4.6 Cell cycle trajectory in untreated condition and its state transition in response to osmotic stress and the α-factor signals. (**a**) The *black* edges show cell cycle state transition in untreated condition. (**b**) The *red* edges illustrate the perturbation from the cell cycle trajectory upon activation of Hog1PP. Moreover, the green edges represent the state transition from the arrest points of osmotic stress to the unperturbed cell cycle trajectory. Osmotic stress cannot take the cell out of its basin, and all perturbations from the limit cycle of the cell cycle can be compensated after the removal of the osmotic stress. (**c**) *Blue* edges illustrate the perturbation from the cell cycle trajectory which is caused by the activity of the α-factor. The α-factor arrests the cell in **fp$_1$**, which is also a fixed point of the cell in osmotic stress condition. The cell which is released from the α-factor will converge to the state $CC3$ (*pink* edge). Hence, the α-factor is not a proper population synchronisation method to study **fp$_1$**. Numbers inside each *circle* are the decimal representation of the corresponding states of cell cycle components

U-basin. This means that time evolution with the matrix O (osmotic) brings these states to the U-basin. Therefore these states can reach the cell cycle trajectory and allow the cell to divide. Hence, a stress can "thaw" frozen states (blue nodes in Fig. 4.4).

A direct description of stresses that move states out of the U-basin is beyond the scope of this chapter. However, some experimental evidence [13] suggests that this prediction might be a biological fact. The experiment regards the behaviour of certain mutated cells in the presence of osmotic stress. This particular mutation affects one key component of the so-called Mitotic Exit Network (MEN). In the absence of osmotic stress these temperature sensitive mutant cells are arrested and cannot exit mitosis, i.e. the mutation modifies the transition network (it deletes edges) so that the U-basin splits into disconnected components. Although the situation discussed above is different, it bears some dynamical similarities. The cell is in a state that cannot reach the fixed point of the U-basin in unstressed conditions; the cells cannot divide and do not form colonies. Remarkably, and consistent with our Boolean model, it was found that these cells do exit mitosis when an osmotic stress is applied. This suggests that a similar observation might be made when the cells are removed from the U-basin by stresses. In this case the stress could bring frozen cells back to the cell cycle trajectory, so that even after removal of the stress the cells would keep dividing.

4.5 Discussion

We present a Boolean model that describes how environmental changes influence the cell cycle dynamics through the entire cell cycle [11]. The model integrates recent experimental findings about the interaction of the osmotic stress response and cell cycle networks across the G1-S-G2-M phases. The finite number of states in the discrete model provides an attractive framework to study the cell cycle dynamics in different environmental conditions.

Osmotic stress kicks the cell away from its cell cycle trajectory. The model predicts that Hog1PP can take the cell into the basin of either of four different fixed points [11]. The state of the cell at the onset of the stress determines the arrest point. Furthermore, the model describes the state transitions under osmotic stress as well as how the cell recovers from osmotic arrest after the removal of the stress. According to the model, all of the four arrest points can reach the cell cycle trajectory when the stress is removed [11].

We discuss that many of the states are to some extent artificial; they are obtained by mathematically generating all possible combinations of states of all constituents. We show that the special relevance of the U-basin (cell cycle basin) is that the cell cycle trajectory lies within this basin, rather than its size. Also, we demonstrate that the size of one of the four cell cycle basins under osmotic stress, is very small compared to the others. We argue that despite its small size, since the experiments to reach that particular basin is known, it is a crucial basin in some experimental settings. We therefore suggest to consider only those states which are *observable*, i.e. can be reached by environmental or experimental perturbations. This suggests a method for reducing the size of state space.

With the developed Boolean model we also study the influence of the α-factor on the dynamical properties of cells [11]. The α-factor is one of the standard methods for population synchronisation of yeast. We show that with this particular method of synchronisation it is not possible to study one of the arrest points of osmotic stress. Hence, other methods of population synchronisation have to be used to study this fixed point and its basin.

Remarkably, our model demonstrates that osmotic stress can move some of the states which are outside of the U-basin back into this basin [11]. This prediction of the model describes the response of MEN mutant cells to osmotic stress. Cells with the MEN mutation cannot go through cell division and are arrested. However, consistent with our Boolean model, these cells divide in the presence of osmotic stress [13]. Hence, osmotic stress takes the MEN cells, the state of which are outside the U-basin, back into this basin. In general, cells that are incapable of forming a colony can be simulated to divide again by osmotic stress. In this case the stress can return them to the cell cycle trajectory, such that they will remain in the U-basin even after the osmotic stress is removed.

References

1. R. Albert, H.G. Othmer, The topology of the regulatory interactions predicts the expression pattern of the segment polarity genes in Drosophila melanogaster. J. Theor. Biol. **223**(1), 1–18 (2003)
2. G. Charvin, C. Oikonomou, E.D. Siggia, F.R. Cross, Origin of irreversibility of cell cycle start in budding yeast. PLoS Biol. **8**(1), e1000284 (2010)
3. K.C. Chen, A. Csikasz-Nagy, B. Gyorffy, J. Val, B. Novak, J.J. Tyson, Kinetic analysis of a molecular model of the budding yeast cell cycle. Mol. Biol. Cell **11**(1), 369–391 (2000)
4. K.C. Chen, L. Calzone, A. Csikasz-nagy, F.R. Cross, B. Novak, J.J. Tyson, Integrative analysis of cell cycle control in budding yeast. Mol. Biol. Cell **15**(August), 3841–3862 (2004)
5. M. Davidich, S. Bornholdt, The transition from differential equations to Boolean networks: a case study in simplifying a regulatory network model. J. Theor. Biol. **255**(3), 269–277 (2008)
6. M.I. Davidich, S. Bornholdt, Boolean network model predicts cell cycle sequence of fission yeast. PLoS ONE **3**(2), e1672 (2008)
7. S. Kauffman, *Origins of Order: Self-organization and Selection in Evolution* (Oxford University Press, Oxford, 1993). Technical monograph
8. E. Klipp, B. Nordlander, R. Kröger, P. Gennemark, S. Hohmann, Integrative model of the response of yeast to osmotic shock. Nat. Biotechnol. **23**(8), 975–982 (2005)
9. F. Li, T. Long, Y. Lu, Q. Ouyang, C. Tang, The yeast cell-cycle network is robustly designed. PNAS **101**(14), 4781–4786 (2004)
10. Y. Okabe, M. Sasai, Stable stochastic dynamics in yeast cell cycle. Biophys. J. **93**(10), 3451–3459 (2007)
11. E. Radmaneshfar, M. Thiel, Recovery from stress: a cell cycle perspective. J. Comp. Int. Sci. **3**(1–2), 33–44 (2012)
12. E. Radmaneshfar, D. Kaloriti, M.C. Gustin, N.A.R Gow, A.J.P Brown, C. Grebogi, M.C. Romano, M. Thiel, From START to FINISH: the influence of osmotic stress on the cell cycle. PLoS ONE **8**(7), e68067 (2013)

References

13. V. Reiser, K.E. D'Aquino, E. Ly-Sha, A. Amon, The stress-activated mitogen-activated protein kinase signaling cascade promotes exit from mitosis. Mol. Biol. Cell **17**(7), 3136–3146 (2006)
14. F. Robert, Discrete iterations: a metric study. Springer Series in Computational Mathematics (1986)
15. J. Saez-Rodriguez, L. Simeoni, J.A. Lindquist, R. Hemenway, U. Bommhardt, B. Arndt, U.-U. Haus, R. Weismantel, E.D. Gilles, S. Klamt, B. Schraven, A logical model provides insights into T cell receptor signaling. PLoS Comput. Biol. **3**(8), 1580–1590 (2007)
16. P.T. Spellman, G. Sherlock, M.Q. Zhang, V.R. Iyer, K. Anders, M.B. Eisen, P.O. Brown, D. Botstein, B. Futcher, Comprehensive identification of cell cycle-regulated genes of the yeast Saccharomyces cerevisiae by microarray hybridization. Mol. Biol. Cell **9**(12), 3273–3297 (1998)
17. M. Tyers, G. Tokiwa, B. Futcher, Comparison of the Saccharomyces cerevisiae G1 cyclins: Cln3 may be an upstream activator of Cln1, Cln2 and other cyclins. EMBO J. **12**(5), 1955–1968 (1993)
18. J.J. Tyson, B. Novak, Regulation of the eukaryotic cell cycle: molecular antagonism, hysteresis, and irreversible transitions. J. Theor. Biol. **210**(2), 249–263 (2001)
19. Y. Zhang, M. Qian, Q. Ouyang, M. Deng, F. Li, C. Tang, Stochastic model of yeast cell-cycle network. Phys. D: Nonlinear Phenom. **219**(1), 35–39 (2006)

Chapter 5
Conclusion

In this work we have studied the mechanisms of cell survival under various environmental conditions from the cell cycle perspective. Proper response to stress is a matter of life or death for a cell. Hence cells have evolved complex and robust networks of molecular interactions which allow them to live in different growth conditions. To understand the intertwined nonlinear and dynamic interactions of the cell cycle in response to environmental stresses we have developed two comprehensive mathematical models. Our models elucidate the emergent dynamics of the cell cycle in response to stresses, and furthermore predict the physiological behaviour of the cell in response to stresses.

First we presented an integrative ODE model which incorporates known and several recently proposed interactions between the stress signalling pathway and cell cycle. Our model illustrates the regulation of the cell from START to FINISH in response to osmotic stress. By considering the complete cell cycle the model elucidates the emergent dynamics of the cell cycle in response to stress which cannot be attributed to single components. The model predicts the possibility of DNA re-replication without cell division upon application of osmotic stress for the cells which are in late S or early G2/M phase at the onset of stress. According to the model, cells stressed at the late G2/M phase exhibit accelerated exit from mitosis and get arrested in the next cell cycle. Furthermore, the model predicts that the cell cycle delay is linearly dependent on the level of stress at the G1, S and early G2/M phase. In contrast, the acceleration in the exit from mitosis does not depend on the level of stress.

Second we presented a Boolean model which is based on the interaction of the osmotic stress and the α-factor with the entire cell cycle network. The finite number of states in this discrete framework allows one to study the dynamical changes of the cell cycle in response to stresses. Our model elucidates the state transition of a cell in the presence of osmotic stress to either of four predicted arrest points. Furthermore, it shows the recovery pathway of the cell from those arrested states when the cell is adapted to the new conditions. This model shows that osmotic stress can cause some of the so called *frozen* cells to divide again. Moreover it shows that although the cell

E. Radmaneshfar, *Mathematical Modelling of the Cell Cycle Stress Response*,
Springer Theses, DOI: 10.1007/978-3-319-00744-1_5,
© Springer International Publishing Switzerland 2014

cycle network is robustly designed, osmotic stress can make cells more fragile. Some of the cell cycle errors which are not lethal in untreated conditions become fatal in the presence of osmotic stress. As in the ODE model, the Boolean model confirms the importance of the state of the cell at the onset of the stress in determination of the arrest point. Furthermore, the model illustrates the dynamical changes of the cell cycle when cells are synchronised by the α-factor. It shows that this method of population synchronisation is not appropriate for the study of cells in the early G1 phase.

In summary, the studies based on two models provide a series of novel predictions for the interactions between the cell cycle and the osmotic stress response, which on the one hand are validated by existing experimental data, and on the other hand, suggest new experiments which are being conducted at the moment of writing this monograph. These experiments will provide us with new data and suggest modifications of the models. The integration of experiments and systems biology deepen our understanding of the cell cycle response to stresses.

Furthermore, at the moment we are working on an important extension of the integrated ODE model, which consists in the inclusion of time fluctuations in the osmotic stress signal. This analysis is key to model more realistic situations, in which cells encounter a fluctuating environment. By changing the frequency and amplitude of the applied osmotic signal it is possible to systematically analyse the effect on the cell cycle progression. Our preliminary results suggest that: (i) cells do not divide if the frequency of the osmotic stress signal is higher than a certain threshold, for sufficiently high dose of stress (1 M NaCl), (ii) the response of the cell cycle to a fluctuating environment does not only depend on the frequency of the signals but also on the amplitude of the osmotic stress signal and (iii) osmotic stress can synchronise a population of cells if the signal is periodic with a frequency similar to the cell cycle one. These predictions provide further insights into the adaptation capabilities of the cell cycle machinery and can be validated using a microfluidic device.

Moreover, the Boolean model can be employed to study the dynamical changes of the cell cycle in response to a randomly fluctuating environment. This approach allows one to analytically study the adaptation of cells to random stress signals. The results illustrate the variation in the robustness of the cell cycle network under this kind of environment.

We have shown that the osmotic stress can cause errors in the process of cell replication and proliferation; therefore it is important to study the influence of stresses on cell cycle progression. Osmotic stress is just one example of the various environmental stresses that cells are exposed to. Many aspects of the cell cycle response to environmental stresses still remain unknown, for example:

(i) The models presented here describe the cell cycle response for a certain range of osmotic stress (0.4 M-1 M NaCl). It has been observed that the response of cells to higher doses of stress is different, which implies that different molecular interactions are involved in this scenario, which are yet to be identified. The models then need to be modified accordingly.

5 Conclusion

(ii) Cells live in environments with many concurrent stresses such as oxidative and nitrosative. Some of the components and links that couple these stresses to the cell cycle are known and others still need to be found. To understand the response of the cell cycle to the presence of these stresses, mathematical models must be developed. Furthermore, these stresses are often present simultaneously and cells are exposed to a combination of them. The response to these stresses are not additive. The cross-talks exist between their signalling pathways which introduce another level of complexity. Therefore a systems biology approach is required to unveil the emergent dynamics as a result of activation of combinatorial stresses.

(iii) Since both the stress response and the cell cycle control network are conserved among eukaryotes, the investigation of the cell cycle response to combinatorial stresses will lead to the discovery of universal mechanisms which can describe the response of different species to various stresses.

These are just some examples of the many future directions of the research presented in this work.

Appendix A
List of Equations, Parameters and Initial Conditions

A.1 Equations

$$\frac{d[mass]}{dt} = \frac{k_g}{1 + k_{dHog1mass}[Hog1PP]}[mass]\left(M_{mass,max} - [mass]\right).\quad\text{(A.1)}$$

$$\frac{d[Cln2]}{dt} = \frac{(kk_{sn2} + kkk_{sn2}[SBF])\,([mass])}{1 + \left(k_{dHog1Cln2}[Hog1PP]\right)^{n_{Hog1Cln2}}} - k_{dn2}[Cln2].\quad\text{(A.2)}$$

$$\begin{aligned}
\frac{d[Clb5]}{dt} = {}& \frac{(kk_{sb5} + kkk_{sb5}[MBF])\,[mass]}{1 + \left(k_{dHog1Clb5}[Hog1PP]\right)^{n_{Hog1Clb5}}} \\
& + (k_{dib5}[C5] + k_{d3c1}[C5P]) + \left(k_{dif5}[F5] + k_{d3f6}[F5P]\right) \\
& + (k_{hdib5}[C5h] + k_{hd3c1}[C5hP]) \\
& - \left(V_{db5} + k_{asb5}[Sic1] + k_{asf5}[Cdc6] + k_{hasb5}[Sic1h]\right)[Clb5].
\end{aligned}$$
$$\text{(A.3)}$$

$$\begin{aligned}
\frac{d[Clb2]}{dt} = {}& \frac{(kk_{sb2} + kkk_{sb2}[Mcm1])\,[mass]}{1 + \left(k_{dHog1Clb2}[Hog1PP]\right)^{n_{Hog1Clb2}}} \\
& + (k_{dib2}[C2] + k_{d3c1}[C2P]) + (k_{hdib2}[C2h] + k_{hd3c1}[C2hP]) \\
& + \left(k_{dif2}[F2] + k_{d3f6}[F2P]\right) \\
& - \left(V_{db2} + k_{asb2}[Sic1] + k_{asf2}[Cdc6] + k_{hasb2}[Sic1h]\right)[Clb2] \\
& + (K_{Mih1}[PClb2] - K_{Swe1}[Clb2]).
\end{aligned}$$
$$\text{(A.4)}$$

E. Radmaneshfar, *Mathematical Modelling of the Cell Cycle Stress Response*,
Springer Theses, DOI: 10.1007/978-3-319-00744-1,
© Springer International Publishing Switzerland 2014

$$\frac{d[Hsl1Hsl7]}{dt} = \frac{kk_{Hsl1Hsl7}[BUD]}{1 + \left(\frac{[Hog1PP]}{J_{Hsl1}}\right)^{n_{Hsl1d}}}$$
$$- kkk_{Hsl1Hsl7}V_{db2}[Hsl1Hsl7] - kkd_{Hsl1Hsl7}[Hsl1Hsl7]. \quad (A.5)$$

$$\frac{d[Sic1h]}{dt} = \frac{kk_{ash}[Sic1][Hog1PP]^{n_{Hog1Sic1}}}{kkk_{ash1}^{n_{Hog1Sic1}} + [Hog1PP]^{n_{Hog1Sic1}}}$$
$$+ (V_{hdb2} + k_{hdib2})[C2h] + (V_{hdb5} + k_{hdib5})[C5h]$$
$$+ (k_{hppc1}[Sic1hP] - k_{h1ppc1}[Sic1h])[Cdc14]$$
$$- (k_{hasb2}[Clb2] + k_{hasb5}[Clb5] + V_{hkpc1})[Sic1h]$$
$$+ k_{hpdib2}[PTrimh] + V_{hpdb2}[PTrimh]$$
$$- k_{hpasb2}[Sic1h][PClb2]. \quad (A.6)$$

$$\frac{d[Sic1hP]}{dt} = V_{hkpc1}[Sic1h] - (k_{hppc1}[Cdc14] + k_{hd3c1})[Sic1hP]$$
$$+ V_{hdb2}[C2hP] + V_{hdb5}[C5hP] + V_{hpdb2}[PTrimhP]$$
$$+ \frac{kk_{ash1}[Sic1P][Hog1PP]^{n_{Hog1Sic1}}}{kkk_{ash1}^{n_{Hog1Sic1}} + [Hog1PP]^{n_{Hog1Sic1}}} - k_{h1ppc1}[Sic1hP][Cdc14].$$
$$(A.7)$$

$$\frac{d[C2]}{dt} = k_{asb2}[Clb2][Sic1] + k_{ppc1}[Cdc14][C2P]$$
$$- (k_{dib2} + V_{db2} + V_{kpc1})[C2]$$
$$- K_{Swe1}[C2] + K_{Mih1}[PTrim] + k_{h1ppc1}[C2h][Cdc14]$$
$$- \frac{kk_{ash1}[C2][Hog1PP]^{n_{Hog1Sic1}}}{kkk_{ash1}^{n_{Hog1Sic1}} + [Hog1PP]^{n_{Hog1Sic1}}}. \quad (A.8)$$

$$\frac{d[C5]}{dt} = k_{asb5}[Clb5][Sic1] + k_{ppc1}[Cdc14][C5P]$$
$$- (k_{dib5} + V_{db5} + V_{kpc1})[C5]$$
$$+ k_{h1ppc1}[C5h][Cdc14] - \frac{kk_{ash1}[C5][Hog1PP]^{n_{Hog1Sic1}}}{kkk_{ash1}^{n_{Hog1Sic1}} + [Hog1PP]^{n_{Hog1Sic1}}}.$$
$$(A.9)$$

Appendix A: List of Equations, Parameters and Initial Conditions   95

$$\frac{d[C2P]}{dt} = V_{kpc1}[C2] - \left(k_{ppc1}[Cdc14] + k_{d3c1} + V_{db2}\right)[C2P]$$
$$- K_{Swe1}[C2P] + K_{Mih1}[PTrimP] + k_{h1ppc1}[C2hP][Cdc14]$$
$$- \frac{kk_{ash1}[C2P][Hog1PP]^{n_{Hog1Sic1}}}{kkk_{ash1}^{n_{Hog1Sic1}} + [Hog1PP]^{n_{Hog1Sic1}}}. \tag{A.10}$$

$$\frac{d[C5P]}{dt} = V_{kpc1}[C5] - \left(k_{ppc1}[Cdc14] + k_{d3c1} + V_{db5}\right)[C5P]$$
$$- \frac{kk_{ash1}[C5P][Hog1PP]^{n_{Hog1Sic1}}}{kkk_{ash1}^{n_{Hog1Sic1}} + [Hog1PP]^{n_{Hog1Sic1}}} + k_{h1ppc1}[C5hP][Cdc14]. \tag{A.11}$$

$$\frac{d[Cdc6]}{dt} = \left(kk_{sf6} + kkk_{sf6}[Swi5] + kkkk_{sf6}[SBF]\right)$$
$$+ \left(V_{db2} + k_{dif2}\right)[F2]$$
$$+ \left(V_{db5} + k_{dif5}\right)[F5] + k_{ppf6}[Cdc14][Cdc6P]$$
$$- \left(k_{asf2}[Clb2] + k_{asf5}[Clb5] + V_{kpf6} + k_{pasf2}[PClb2]\right)[Cdc6]$$
$$+ \left(V_{pdb2} + k_{pdif2}\right)[PF2]. \tag{A.12}$$

$$\frac{d[Cdc6P]}{dt} = V_{kpf6}[Cdc6] - \left(k_{ppf6}[Cdc14] + k_{d3f6}\right)[Cdc6P]$$
$$+ V_{db2}[F2P] + V_{db5}[F5P] + V_{pdb2}[PF2P]. \tag{A.13}$$

$$\frac{d[F2]}{dt} = k_{asf2}[Clb2][Cdc6] + k_{ppf6}[Cdc14][F2P]$$
$$- \left(k_{dif2} + V_{db2} + V_{kpf6}\right)[F2] - K_{Swe1}[F2] + K_{Mih1}[PF2]. \tag{A.14}$$

$$\frac{d[F5]}{dt} = k_{asf5}[Clb5][Cdc6] + k_{ppf6}[Cdc14][F5P]$$
$$- \left(k_{dif5} + V_{db5} + V_{kpf6}\right)[F5]. \tag{A.15}$$

$$\frac{d[F2P]}{dt} = V_{kpf6}[F2] - \left(k_{ppf6}[Cdc14] + k_{d3f6} + V_{db2}\right)[F2P]$$
$$- K_{Swe1}[F2P] + K_{Mih1}[PF2P]. \tag{A.16}$$

$$\frac{d[F5P]}{dt} = V_{kpf6}[F5] - \left(k_{ppf6}[Cdc14] + k_{d3f6} + V_{db5}\right)[F5P]. \tag{A.17}$$

$$\frac{d[Swi5_T]}{dt} = kk_{sswi} + kkk_{sswi}[Mcm1] - k_{dswi}[Swi5_T]. \tag{A.18}$$

$$\frac{d[Swi5]}{dt} = kk_{sswi} + kkk_{sswi}[Mcm1].$$
$$+ k_{aswi}[Cdc14]\,([Swi5_T] - [Swi5])$$
$$- (k_{dswi} + k_{iswi}[Clb2])\,[Swi5]. \tag{A.19}$$

$$\frac{d[APC_P]}{dt} = \frac{k_{aapc}[Clb2]\,(1 - [APC_P])}{J_{aapc} + 1 - [APC_P]} - \frac{k_{iapc}[APC_P]}{J_{iapc} + [APC_P]}. \tag{A.20}$$

$$\frac{d[Cdc20_T]}{dt} = kk_{s20} + kkk_{s20}[Mcm1] - k_{d20}[Cdc20_T]. \tag{A.21}$$

$$\frac{d[Cdc20_A]}{dt} = (kk_{a20} + kkk_{a20}[APC_P])\,([Cdc20_T] - [Cdc20_A])$$
$$- (k_{mad2} + k_{d20})\,[Cdc20_A]. \tag{A.22}$$

$$\frac{d[Cdh1]}{dt} = k_{scdh} - k_{dcdh}[Cdh1]$$
$$+ \frac{V_{acdh}\,([Cdh1_T] - [Cdh1])}{J_{acdh} + [Cdh1_T] - [Cdh1]} - \frac{V_{icdh}[Cdh1]}{J_{icdh} + [Cdh1]}. \tag{A.23}$$

$$\frac{d[Cdh1_T]}{dt} = k_{scdh} - k_{dcdh}[Cdh1_T]. \tag{A.24}$$

$$\frac{d[Tem1]}{dt} = \frac{k_{lte1}\,([Tem1_T] - [Tem1])}{J_{atem} + Tem1_T - Tem1} - \frac{k_{bub2}[Tem1]}{J_{item} + [Tem1]}. \tag{A.25}$$

$$\frac{d[Cdc15]}{dt} = -k_{i15}[Cdc15]$$
$$+ (kk_{a15}\,([Tem1_T] - [Tem1]))\,([Cdc15_T] - [Cdc15])$$
$$+ (kkk_{a15}[Tem1] + kkkk_{a15}[Cdc14])\,([Cdc15_T] - [Cdc15]). \tag{A.26}$$

Appendix A: List of Equations, Parameters and Initial Conditions

$$\frac{d[Cdc14_T]}{dt} = k_{s14} - k_{d14}[Cdc14_T]. \tag{A.27}$$

$$\frac{d[Cdc14]}{dt} = k_{s14} - k_{d14}[Cdc14] + k_{dnet}\,([RENT] + [RENTP]) \\ + k_{dirent}[RENT] + k_{direntp}[RENTP] \\ - \left(k_{asrent}[Net1] + k_{asrentp}[Net1P]\right)[Cdc14]. \tag{A.28}$$

$$\frac{d[Net1_T]}{dt} = k_{snet} - k_{dnet}[Net1_T]. \tag{A.29}$$

$$\frac{d[Net1]}{dt} = k_{snet} - k_{dnet}[Net1] + k_{d14}[RENT] + k_{dirent}[RENT] \\ - k_{asrent}[Cdc14][Net1] + V_{ppnet}[Net1P] - V_{kpnet}[Net1]. \tag{A.30}$$

$$\frac{d[RENT]}{dt} = -\,(k_{d14} + k_{dnet})\,[RENT] \\ - k_{dirent}[RENT] + k_{asrent}[Cdc14][Net1] \\ - V_{kpnet}[RENT] + V_{ppnet}[RENTP]. \tag{A.31}$$

$$\frac{d[PPX]}{dt} = k_{sppx} - V_{dppx}[PPX]. \tag{A.32}$$

$$\frac{d[Pds1]}{dt} = kk_{spds} + kkk_{s1pds}[SBF] + kkk_{s2pds}[Mcm1] \\ + k_{diesp}[PE] - \left(V_{dpds} + k_{asesp}[Esp1]\right)[Pds1]. \tag{A.33}$$

$$\frac{d[Esp1]}{dt} = -k_{asesp}[Pds1][Esp1] + \left(k_{diesp} + V_{dpds}\right)[PE]. \tag{A.34}$$

$$\frac{d[ORI]}{dt} = k_{sori}\,(\varepsilon_{orib5}[Clb5] + \varepsilon_{orib2}[Clb2]) - k_{dori}[ORI]. \tag{A.35}$$

$$\frac{d[BUD]}{dt} = k_{sbud}\,(\varepsilon_{budn2}[Cln2] + \varepsilon_{budn3}[Cln3] + \varepsilon_{budb5}[Clb5]) \\ - k_{dbud}[BUD] \tag{A.36}$$

$$\frac{d[SPN]}{dt} = k_{sspn}\frac{[Clb2]}{Jspn + [Clb2]} - k_{dspn}[SPN]. \tag{A.37}$$

$$\frac{d[Swe1]}{dt} = k_{sswe}[SBF] + k_{ssweC} - \frac{k_{hsl1}[Hsl1Hsl7][Swe1]}{J_{iwee} + [Swe1]}$$
$$+ k_{hsl1r}[Swe1M] - \frac{V_{iwee}[Clb2][Swe1]}{J_{iwee} + [Swe1]}$$
$$+ \frac{V_{awee}[Swe1P]}{J_{awee} + [Swe1P]} - kk_{dswe}[Swe1]. \tag{A.38}$$

$$\frac{d[Swe1P]}{dt} = -\frac{k_{hsl1}[Hsl1Hsl7][Swe1P]}{J_{iwee} + [Swe1P]} + k_{hsl1r}[Swe1MP]$$
$$- \frac{V_{awee}[Swe1P]}{J_{awee} + [Swe1P]} - kk_{dswe}[Swe1P]$$
$$+ \frac{V_{iwee}[Clb2][Swe1]}{J_{iwee} + [Swe1]}. \tag{A.39}$$

$$\frac{d[Swe1MP]}{dt} = \frac{k_{hsl1}[Hsl1Hsl7][Swe1P]}{J_{iwee} + [Swe1P]} - k_{hsl1r}[Swe1MP]$$
$$- \frac{V_{awee}[Swe1MP]}{J_{awee} + [Swe1MP]} + \frac{V_{iwee}[Clb2][Swe1M]}{J_{iwee} + [Swe1M]}$$
$$- kk_{dswe}[Swe1MP]. \tag{A.40}$$

$$\frac{d[Swe1M]}{dt} = \frac{k_{hsl1}[Hsl1Hsl7][Swe1]}{J_iwee + [Swe1]} - k_{hsl1r}[Swe1M]$$
$$- \frac{V_{iwee}[Clb2][Swe1M]}{J_{iwee} + [Swe1M]}$$
$$+ \frac{V_{awee}[Swe1MP]}{J_{awee} + [Swe1MP]} - kk_{dswe}[Swe1M]. \tag{A.41}$$

$$\frac{d[PClb2]}{dt} = K_{Swe1}[Clb2] - K_{Mih1}[PClb2] + k_{pd3c1}[PTrimP]$$
$$+ k_{pdib2}[PTrim] + k_{pd3f6}[PF2P] + k_{pdif2}[PF2]$$
$$- V_{pdb2}[PClb2] - k_{pasb2}[Sic1][PClb2]$$
$$+ k_{hpd3c1}[PTrimhP] + k_{hpdib2}[PTrimh]$$
$$- k_{pasf2}[Cdc6][PClb2] - k_{hpasb2}[PClb2][Sic1h]. \tag{A.42}$$

Appendix A: List of Equations, Parameters and Initial Conditions

$$\frac{d[PTrim]}{dt} = k_{pasb2}[Sic1][PClb2] - k_{pdib2}[PTrim] + K_{Swe1}[C2]$$
$$+ \left(k_{pppc1}[PTrimP] + k_{h1ppc1}[PTrimh]\right)[Cdc14]$$
$$- V_{pkpc1}[PTrim] - K_{Mih1}[PTrim] - V_{pdb2}[PTrim]$$
$$- \frac{kk_{ash1}[PTrim][Hog1PP]^{n_{Hog11Sic1}}}{kkk_{ash1}^{n_{Hog1Sic1}} + [Hog1PP]^{n_{Hog1Sic1}}}. \tag{A.43}$$

$$\frac{d[Mih1]}{dt} = \frac{Va_{mih}[Clb2]([Mih1_T] - [Mih1])}{Ja_{mih} + [Mih1_T] - [Mih1]} - \frac{Vi_{mih}[Mih1]}{Ji_{mih} + [Mih1]}. \tag{A.44}$$

$$\frac{d[PTrimP]}{dt} = -k_{pd3c1}[PTrimP] - k_{pppc1}[Cdc14][PTrimP]$$
$$+ V_{pkpc1}[PTrim] - V_{pdb2}[PTrimP]$$
$$- K_{Mih1}[PTrimP] + K_{Swe1}[C2P]$$
$$- \frac{kk_{ash1}[PTrimP][Hog1PP]^{n_{Hog1Sic1}}}{kkk_{ash1}^{n_{Hog1Sic1}} + [Hog1PP]^{n_{Hog1Sic1}}}$$
$$+ k_{h1ppc1}[Cdc14][PTrimhP]. \tag{A.45}$$

$$\frac{d[PF2]}{dt} = -k_{pdif2}[PF2] + k_{pasf2}[PClb2][Cdc6]$$
$$+ k_{pppf6}[Cdc14][PF2P]$$
$$- V_{pdb2}[PF2] - V_{pkpf6}[PF2]$$
$$+ K_{Swe1}[F2] - K_{Mih1}[PF2]. \tag{A.46}$$

$$\frac{d[PF2P]}{dt} = V_{pkpf6}[PF2] - k_{pppf6}[Cdc14][PF2P]$$
$$- k_{pd3f6}[PF2P] - V_{pdb2}[PF2P]$$
$$+ K_{Swe1}[F2P] - K_{Mih1}[PF2P]. \tag{A.47}$$

$$\frac{d[C5h]}{dt} = k_{hasb5}[Clb5][Sic1h] + k_{hppc1}[Cdc14][C5hP]$$
$$- \left(k_{hdib5} + V_{hdb5} + V_{hkpc1}\right)[C5h]$$
$$+ \frac{kk_{ash1}[C5][Hog1PP]^{n_{Hog1Sic1}}}{kkk_{ash1}^{n_{Hog1Sic1}} + [Hog1PP]^{n_{Hog1Sic1}}}$$
$$- k_{h1ppc1}[C5h][Cdc14]. \tag{A.48}$$

$$\frac{d[C2h]}{dt} = k_{hasb2}[Clb2][Sic1h] + k_{hppc1}[Cdc14][C2hP]$$
$$- \left(k_{hdib2} + V_{hdb2} + V_{hkpc1}\right)[C2h]$$
$$- K_{Swe1}[C2h] + K_{Mih1}[PTrimh] - k_{h1ppc1}[C2h][Cdc14]$$
$$+ \frac{kk_{ash1}[C2][Hog1PP]^{n_{Hog1Sic1}}}{kkk_{ash1}^{n_{Hog1Sic1}} + [Hog1PP]^{n_{Hog1Sic1}}}. \tag{A.49}$$

$$\frac{d[C5hP]}{dt} = V_{hkpc1}[C5h] - k_{h1ppc1}[C5hP][Cdc14]$$
$$- \left(k_{hppc1}[Cdc14] + k_{hd3c1} + V_{hdb5}\right)[C5hP]$$
$$+ \frac{kk_{ash1}[C5P][Hog1PP]^{n_{Hog1Sic1}}}{kkk_{ash1}^{n_{Hog1Sic1}} + [Hog1PP]^{n_{Hog1Sic1}}}. \tag{A.50}$$

$$\frac{d[Sic1P]}{dt} = V_{kpc1}[Sic1] - \left(k_{ppc1}[Cdc14] + k_{d3c1}\right)[Sic1P]$$
$$+ V_{db2}[C2P] + V_{db5}[C5P] + V_{pdb2}[PTrimP]$$
$$- \frac{kk_{ash1}[Sic1P][Hog1PP]^{n_{Hog1Sic1}}}{kkk_{ash1}^{n_{Hog1Sic1}} + [Hog1PP]^{n_{Hog1Sic1}}} + k_{h1ppc1}[Sic1hP][Cdc14]. \tag{A.51}$$

$$\frac{d[C2hP]}{dt} = V_{hkpc1}[C2h] - k_{h1ppc1}[C2hP][Cdc14]$$
$$- \left(k_{hppc1}[Cdc14] + k_{hd3c1} + V_{hdb2}\right)[C2hP]$$
$$- K_{Swe1}[C2hP] + K_{Mih1}[PTrimhP]$$
$$+ \frac{kk_{ash1}[C2P][Hog1PP]^{n_{Hog1Sic1}}}{kkk_{ash1}^{n_{Hog1Sic1}} + [Hog1PP]^{n_{Hog1Sic1}}}. \tag{A.52}$$

$$\frac{d[PTrimh]}{dt} = k_{hpasb2}[Sic1h][PClb2] - k_{hpdib2}[PTrimh]$$
$$+ k_{hpppc1}[Cdc14][PTrimhP] - k_{h1ppc1}[Cdc14][PTrimh]$$
$$- V_{hpdb2}[PTrimh] - V_{hpkpc1}[PTrimh]$$
$$+ K_{Swe1}[C2h] - K_{Mih1}[PTrimh]$$
$$+ \frac{kk_{ash1}[PTrim][Hog1PP]^{n_{Hog1Sic1}}}{kkk_{ash1}^{n_{Hog1Sic1}} + [Hog1PP]^{n_{Hog1Sic1}}}. \tag{A.53}$$

Appendix A: List of Equations, Parameters and Initial Conditions

$$
\begin{aligned}
\frac{d[PTrimhP]}{dt} = {} & -k_{hpd3c1}[PTrimhP] - k_{hpppc1}[Cdc14][PTrimhP] \\
& + V_{hpkpc1}[PTrimh] - V_{hpdb2}[PTrimhP] \\
& - K_{Mih1}[PTrimhP] + K_{Swe1}[C2hP] \\
& - k_{h1ppc1}[Cdc14][PTrimhP] \\
& + \frac{kk_{ash1}[PTrimP][Hog1PP]^{n_{Hog1Sic1}}}{kkk_{ash1}^{n_{Hog1Sic1}} + [Hog1PP]^{n_{Hog1Sic1}}}.
\end{aligned}
\tag{A.54}
$$

$$
\begin{aligned}
\frac{d[Sic1]}{dt} = {} & (kk_{sc1} + kkk_{sc1}[Swi5]) + (V_{db2} + k_{dib2})\,[C2] \\
& + (V_{db5} + k_{dib5})\,[C5] + k_{ppc1}[Cdc14][Sic1P] \\
& - \left(k_{asb2}[Clb2] + k_{asb5}[Clb5] + V_{kpc1}\right)[Sic1] \\
& - \frac{kk_{ash}[Sic1][Hog1PP]^{n_{Hog1Sic1}}}{kkk_{ash1}^{n_{Hog1Sic1}} + [Hog1PP]^{n_{Hog1Sic1}}} \\
& - k_{pasb2}[Sic1][PClb2] + k_{h1ppc1}[Sic1h][Cdc14] \\
& + \left(V_{pdb2} + k_{pdib2}\right)[PTrim].
\end{aligned}
\tag{A.55}
$$

$$
[Bck2] = B_0[mass].
\tag{A.56}
$$

$$
\begin{aligned}
[Clb5_T] = {} & [Clb5] + [C5] + [C5P] \\
& + [F5] + [F5P] + [C5h] + [C5hP].
\end{aligned}
\tag{A.57}
$$

$$
\begin{aligned}
[Clb2_T] = {} & [Clb2] + [C2] + [C2P] + [F2] + [F2P] \\
& + [PClb2] + [PTrim] + [PTrimP] \\
& + [PF2] + [PF2P] + [C2h] + [C2hP] \\
& + [PTrimh] + [PTrimhP].
\end{aligned}
\tag{A.58}
$$

$$
\begin{aligned}
[Sic1_T] = {} & [Sic1] + [Sic1P] + [C2] + [C2P] \\
& + [C5] + [C5P] + [PTrim] + [PTrimP] \\
& + [Sic1h] + [Sic1hP] + [C2h] + [C2hP] \\
& + [C5h] + [C5hP] + [PTrimh] + [PTrimhP].
\end{aligned}
\tag{A.59}
$$

$$
\begin{aligned}
[Cdc6_T] = {} & [Cdc6] + [Cdc6P] + [F2] + [F2P] \\
& + [F5] + [F5P] + [PF2] + [PF2P].
\end{aligned}
\tag{A.60}
$$

$$[CKI_T] = [Sic1_T] + [Cdc6_T].\tag{A.61}$$

$$[RENTP] = [Cdc14_T] - [RENT] - [Cdc14].\tag{A.62}$$

$$[Net1P] = [Net1_T] - [Net1] - [Cdc14_T] + [Cdc14].\tag{A.63}$$

$$[PE] = [Esp1_T] - [Esp1].\tag{A.64}$$

$$V_{db5} = kk_{db5} + kkk_{db5}[Cdc20_A].\tag{A.65}$$

$$V_{db2} = kk_{db2} + kkk_{db2}[Cdh1] + k_{db2p}[Cdc20_A].\tag{A.66}$$

$$V_{asbf} = k_{asbf}\left(\varepsilon_{sbfn2}[Cln2] + \varepsilon_{sbfn3}\left([Cln3] + [Bck2]\right)\right) \\ + k_{asbf}\left(\varepsilon_{sbfb5}[Clb5]\right).\tag{A.67}$$

$$V_{isbf} = kk_{isbf} + kkk_{isbf}[Clb2].\tag{A.68}$$

$$V_{acdh} = kk_{acdh} + kkk_{acdh}[Cdc14].\tag{A.69}$$

$$V_{icdh} = kk_{icdh} \\ + kkk_{icdh}\left(\varepsilon_{cdhn3}[Cln3] + \varepsilon_{cdhn2}[Cln2] + \varepsilon_{cdhb2}[Clb2]\right. \\ \left. + \varepsilon_{cdhb5}[Clb5]\right).\tag{A.70}$$

$$V_{ppnet} = kk_{ppnet} + kkk_{ppnet}[PPX].\tag{A.71}$$

$$V_{kpnet} = \left(kk_{kpnet} + kkk_{kpnet}[Cdc15]\right)[mass].\tag{A.72}$$

$$V_{dppx} = kk_{dppx} + kkk_{dppx}\left(J_{20ppx} + [Cdc20_A]\right)\frac{J_{pds}}{J_{pds} + [Pds1]}.\tag{A.73}$$

$$V_{dpds} = kk_{d1pds} + kkk_{d2pds}[Cdc20_A] + kkk_{d3pds}[Cdh1].\tag{A.74}$$

$$K_{Swe1} = kk_{swe}[Swe1] + kkk_{swe}[Swe1M] + kkkk_{swe}[Swe1P].\tag{A.75}$$

$$K_{Mih1} = kk_{mih1}[Mih1] + kkk_{mih1}\left([Mih1_T] - [Mih1]\right).\tag{A.76}$$

$$[MBF] = G\left(V_{asbf}, V_{isbf}, J_{asbf}, J_{isbf}\right).\tag{A.77}$$

Appendix A: List of Equations, Parameters and Initial Conditions   103

$$[SBF] = G\left(V_{asbf}, V_{isbf}, J_{asbf}, J_{isbf}\right).$$ (A.78)

$$[Mcm1] = G\left(k_{amcm}[Clb2], k_{imcm}, J_{amcm}, J_{imcm}\right).$$ (A.79)

$$V_{hkpc1} = k_{hkpc1} V_{kpc1}.$$ (A.80)

$$V_{hpkpc1} = k_{hpkpc1} V_{kpc1}.$$ (A.81)

$$V_{pdb2} = V_{db2}.$$ (A.82)

$$V_{pkpf6} = V_{kpf6}.$$ (A.83)

$$V_{kpc1} = k_{d1c1}$$
$$+ k_{d2c1} \frac{\varepsilon_{c1n3}[Cln3] + \varepsilon_{c1k2}[Bck2] + \varepsilon_{c1n2}[Cln2] + \varepsilon_{c1b5}[Clb5] + \varepsilon_{c1b2}[Clb2]}{J_{d2c1} + [Sic1_T]}.$$
(A.84)

$$V_{kpf6} = k_{d1f6}$$
$$+ k_{d2f6} \frac{\varepsilon_{f6n3}[Cln3] + \varepsilon_{f6k2}[Bck2] + \varepsilon_{f6n2}[Cln2] + \varepsilon_{f6b5}[Clb5] + \varepsilon_{f6b2}[Clb2]}{J_{d2f6} + [Cdc6_T]}.$$
(A.85)

$$G\left(V_a, V_i, J_a, J_i\right) =$$
$$\frac{2 J_i V_a}{V_i - V_a + J_a V_i + J_i V_a + \sqrt{(V_i - V_a + J_a V_i + J_i V_a)^2 - 4(V_i - V_a) J_i V_a}}.$$
(A.86)

$$V_{hdb5} = V_{db5}.$$ (A.87)

$$V_{hdb2} = V_{db2}.$$ (A.88)

$$V_{hpdb2} = V_{db2}.$$ (A.89)

$$V_{pkpc1} = V_{kpc1}.$$ (A.90)

A.2 Initial Conditions and Parameters

Table A.1 Initial conditions for a newborn, wild-type daughter cell in arbitrary units

$[mass] = 2.201$	$[Cln2] = 5.069 \times 10^{-2}$	$[Clb5] = 7.330 \times 10^{-2}$	$[Clb2] = 3.008 \times 10^{-1}$
$[Sic1] = 1.123 \times 10^{-2}$	$[Sic1P] = 7.408 \times 10^{-3}$	$[C2] = 2.086 \times 10^{-1}$	$[C5] = 4.810 \times 10-2$
$[C2P] = 3.994 \times 10-2$	$[C5P] = 9.154 \times 10^{-3}$	$[Cdc6] = 5.911 \times 10^{-2}$	$[Cdc6P] = 1.808 \times 10^{-2}$
$[F2] = 2.154 \times 10^{-1}$	$[F5] = 5.007 \times 10^{-5}$	$[F2P] = 4.681 \times 10^{-2}$	$[F5P] = 1.055 \times 10^{-5}$
$[Swi5_T] = 9.870 \times 10^{-1}$	$[Swi5] = 9.532 \times 10^{-1}$	$[APC_P] = 9.676 \times 10^{-2}$	$[Cdc20_T] = 1.965$
$[Cdc20_A] = 4.241 \times 10^{-1}$	$[Cdh1_T] = 1$	$[Cdh1] = 6.675 \times 10^{-1}$	$[Tem1] = 9.778 \times 10^{-1}$
$[Cdc15] = 6.618 \times 10^{-1}$	$[Cdc14_T] = 2$	$[Cdc14] = 4.370 \times 10^{-1}$	$[Net1_T] = 2.8$
$[Net1] = 2.000 \times 10^{-2}$	$[RENT] = 1.014$	$[PPX] = 1.381 \times 10^{-1}$	$[Pds1] = 3.284 \times 10^{-2}$
$[Esp1] = 2.584 \times 10^{-1}$	$[ORI] = 9 \times 10^{-4}$	$[BUD] = 8.5 \times 10^{-3}$	$[SPN] = 3.05 \times 10^{-2}$
$[Hsl1Hsl7] = 8.236 \times 10^{-2}$	$[Swe1P] = 6.066 \times 10^{-4}$	$[Mih1] = 8.625 \times 10^{-1}$	$[PClb2] = 5.246 \times 10^{-4}$
$[PTrim] = 3.789 \times 10^{-4}$	$[PTrimP] = 7.324 \times 10^{-5}$	$[PF2] = 3.902 \times 10^{-4}$	$[PF2P] = 8.528 \times 10^{-5}$
$[Sic1h] = 1.262 \times 10^{-6}$	$[Sic1hP] = 1.096 \times 10^{-5}$	$[C5h] = 7.165 \times 10^{-6}$	$[C2h] = 3.111 \times 10^{-5}$
$[C5hP] = 7.489 \times 10^{-6}$	$[C2hP] = 3.252 \times 10^{-5}$	$[PTrimh] = 6.318 \times 10^{-8}$	$[PTrimhP] = 1.075 \times 10^{-7}$
$[Swe1MP] = 1.829 \times 10^{-2}$	$[Swe1M] = 6.223 \times 10^{-1}$	$[Swe1] = 1.004 \times 10^{-3}$	

Appendix A: List of Equations, Parameters and Initial Conditions

Table A.2 Parameters: the dimension of all parameters that start with a lower case k is min^{-1}

Parameters	References	Parameters	References	Parameters	References
$k_g = 0.007702$	(Chen 2000)	$kk_{sn2} = 0$	(Cross 2002)	$kkk_{sn2} = 0.15$	(Cross 2002)
$k_{dn2} = 0.12$	(Salama 1994)	$kk_{sb5} = 0.0008$	(Cross 2002)	$kkk_{sb5} = 0.005$	(Cross 2002)
$kkk_{db2} = 0.4$	(Cross 2002)	$k_{d3c1} = 1$	(Cross 2002)	$k_{d2f6} = 1$	(Cross 2002)
$kkk_{sswi} = 0.08$	(Chen 2004)	$kk_{db5} = 0.01$	(Cross 2002)	$kkk_{db5} = 0.16$	(Chen 2004)
$kk_{sb2} = 0.001$	(Cross 2002)	$kkk_{sb2} = 0.04$	(Cross 2002)	$kk_{db2} = 0.003$	(Cross 2002)
$k_{db2p} = 0.15$	(Cross 2002)	$kk_{sc1} = 0.012$	(Chen 2004)	$kkk_{sc1} = 0.12$	(Chen 2004)
$k_{d1c1} = 0.01$	(Chen 2004)	$k_{d2c1} = 1$	(Chen 2004)	$k_{ppc1} = 4$	(Chen 2004)
$kk_{sf6} = 0.024$	(Chen 2004)	$kkk_{sf6} = 0.12$	(Chen 2004)	$kkkk_{sf6} = 0.004$	(Chen 2004)
$k_{d1f6} = 0.01$	(Chen 2004)	$k_{d3f6} = 1$	(Chen 2004)	$k_{ppf6} = 4$	(Chen 2004)
$k_{asb5} = 50$	(Chen 2000)	$k_{dib5} = 0.06$	(Chen 2000)	$k_{asf5} = 0.01$	(Chen 2000)
$k_{asb2} = 50$	(Chen 2000)	$k_{dib2} = 0.05$	(Chen 2000)	$k_{asf2} = 15$	(Chen 2000)
$k_{dif2} = 0.5$	(Chen 2000)	$kk_{sswi} = 0.005$	(Chen 2000)	$k_{dif5} = 0.01$	(Chen 2004)
$kk_{s20} = 0.006$	(Chen 2004)	$k_{dcdh} = 0.01$	(Chen 2004)	$k_{d14} = 0.1$	(Chen 2004)
$k_{i15} = 0.5$	(Chen 2004)	$k_{dswi} = 0.08$	(Chen 2004)	$k_{aswi} = 2$	(Chen 2004)
$k_{iswi} = 0.05$	(Chen 2004)	$k_{aapc} = 0.1$	(Chen 2004)	$k_{iapc} = 0.15$	(Chen 2004)
$kkk_{s20} = 0.6$	(Chen 2004)	$k_{d20} = 0.3$	(Chen 2004)	$kk_{a20} = 0.05$	(Chen 2004)
$kkk_{a20} = 0.2$	(Chen 2004)	$k_{acdh} = 0.01$	(Chen 2004)	$kk_{acdh} = 0.01$	(Chen 2004)
$kkk_{acdh} = 0.8$	(Chen 2004)	$kk_{icdh} = 0.001$	(Chen 2004)	$kkk_{icdh} = 0.08$	(Chen 2004)
$k_{s14} = 0.2$	(Chen 2004)	$k_{snet} = 0.084$	(Chen 2004)	$k_{dnet} = 0.03$	(Chen 2004)
$kk_{a15} = 0.002$	(Chen 2004)	$kkk_{a15} = 1$	(Chen 2004)	$kkkk_{a15} = 0.001$	(Chen 2004)
$kk_{ppnet} = 0.05$	(Chen 2004)	$kkk_{ppnet} = 3$	(Chen 2004)	$kk_{kpnet} = 0.01$	(Chen 2004)
$kkk_{kpnet} = 0.6$	(Chen 2004)	$k_{asrent} = 200$	(Chen 2004)	$k_{asrentp} = 1$	(Chen 2004)
$kk_{spds} = 0$	(Chen 2004)	$k_{asesp} = 50$	(Chen 2004)	$k_{sspn} = 0.1$	(Chen 2004)
$k_{imcm} = 0.15$	(Chen 2004)	$k_{dirent} = 1$	(Chen 2004)	$k_{direntp} = 2$	(Chen 2004)
$k_{sppx} = 0.1$	(Chen 2004)	$kk_{dppx} = 0.17$	(Chen 2004)	$kkk_{dppx} = 2$	(Chen 2004)
$kkk_{s1pds} = 0.03$	(Chen 2004)	$kkk_{s2pds} = 0.055$	(Chen 2004)	$kk_{d1pds} = 0.01$	(Chen 2004)
$kkk_{d2pds} = 0.2$	(Chen 2004)	$kkk_{d3pds} = 0.04$	(Chen 2004)	$k_{diesp} = 0.5$	(Chen 2004)
$k_{sori} = 2$	(Su 1995)	$k_{dori} = 0.06$	(Su 1995)	$k_{sbud} = 0.2$	(Zachariae 1999)
$k_{dbud} = 0.06$	(Zachariae 1999)	$k_{dspn} = 0.06$	(Su 1995)	$k_{asbf} = 0.38$	(Dirick 1995)
$kk_{sbf} = 0.642$	(Dirick 1995)	$kkk_{sbf} = 8$	(Amon 1993)	$k_{amcm} = 1$	(Chen 2004)
$\varepsilon_{sbfn2} = 2$	(Dirick 1995)	$\varepsilon_{sbfn3} = 10$	(Epstein 1994)	$\varepsilon_{sbfb5} = 2$	(Schwob 1994)
$\varepsilon_{c1n3} = 0.3$	(Epstein 1994)	$\varepsilon_{c1n2} = 0.06$	(Dirick 1995)	$\varepsilon_{f6b5} = 0.1$	(Schwob 1994)
$J_{icdh} = 0.03$	(Chen 2004)	$J_{imcm} = 0.1$	(Chen 2004)	$K_{ez2} = 0.2$	(Su 1995)
$M_{mass,max} = 2.5$	this study	$\varepsilon_{c1k2} = 0.03$	(Epstein 1994)	$\varepsilon_{orib5} = 0.9$	(Epstein 1994)
$\varepsilon_{orib2} = 0.45$	(Epstein 1994)	$n_{Hog1Clb2} = 8$	this study	$k_{ash1} = 0.25$	(Chen 2004)
$\varepsilon_{c1b5} = 0.1$	(Epstein 1994)	$\varepsilon_{c1b2} = 0.45$	(Epstein 1994)	$\varepsilon_{f6n3} = 0.3$	(Epstein 1994)
$\varepsilon_{f6n2} = 0.06$	(Epstein 1994)	$\varepsilon_{f6k2} = 0.03$	(Epstein 1994)	$\varepsilon_{f6b2} = 0.55$	(Epstein 1994)
$\varepsilon_{cdhn3} = 0.25$	(Epstein 1994)	$\varepsilon_{cdhn2} = 0.4$	(Epstein 1994)	$\varepsilon_{cdhb5} = 8$	(Epstein 1994)
$\varepsilon_{cdhb2} = 1.2$	(Epstein 1994)	$\varepsilon_{budn3} = 0.05$	(Epstein 1994)	$\varepsilon_{budn2} = 0.25$	(Epstein 1994)
$\varepsilon_{budb5} = 1$	(Epstein 1994)	$C_0 = 0.4$	this study	$D_{n3} = 1$	this study
$J_{d2c1} = 0.05$	(Chen 2004)	$J_{d2f6} = 0.05$	(Chen 2004)	$J_{aapc} = 0.1$	(Chen 2004)
$J_{iapc} = 0.1$	(Chen 2004)	$J_{acdh} = 0.03$	(Chen 2004)	$J_{atem} = 0.1$	(Chen 2004)
$J_{item} = 0.1$	(Chen 2004)	$J_{asbf} = 0.01$	(Ciliberto 2003)	$J_{isbf} = 0.01$	(Ciliberto 2003)
$J_{amcm} = 0.1$	(Ciliberto 2003)	$J_{spn} = 0.14$	(Chen 2004)	$J_{20ppx} = 0.15$	(Chen 2004)
$J_{pds} = 0.04$	(Chen 2004)	$K_{ez} = 0.3$	(Zachariae 1999)	$k_{pd3c1} = k_{d3c1}$	this study
$k_{pdib2} = k_{dib2}$	this study	$k_{pd3f6} = k_{d3f6}$	this study	$k_{pdif2} = k_{dif2}$	this study
$k_{pasb2} = k_{asb2}$	this study	$k_{pasf2} = k_{asf2}$	this study	$k_{pppc1} = k_{ppc1}$	this study
$k_{ppf6} = k_{ppf6}$	this study	$k_{npd3c1} = 0.01$	this study, (Escoté 2004)	$k_{hpdib2} = k_{dib2}$	this study
$k_{hpasb2} = k_{asb2}$	this study	$k_{hd3c1} = 0.01$	this study, (Escoté 2004)	$k_{hpppc1} = k_{ppc1}$	this study
$k_{hdib5} = k_{dib5}$	this study	$k_{hasb5} = k_{asb5}$	this study	$k_{hdib2} = k_{dib2}$	this study
$k_{hasb2} = k_{asb2}$	this study	$k_{hppc1} = k_{ppc1}$	this study	$k_{h1ppc1} = 0.68$	this study
$n_{Hog1S1c1} = 2$	this study	$k_{dHog1Clb2} = 20$	this study	$k_{dHog1Clb5} = 0.01$	this study
$J_{Hsl1} = 0.012$	this study	$n_{Hsl1d} = 8$	this study	$k_{sswe} = 0.006$	(Ciliberto 2003)
$k_{ssweC} = 0$	(Ciliberto 2003)	$k_{hsl1} = 0.5$	this study	$k_{hsl1r} = 0.01$	(Ciliberto 2003)
$V_{awee} = 0.3$	(Ciliberto 2003)	$V_{iwee} = .2$	(Ciliberto 2003)	$kk_{dswe} = 0.007$	(Ciliberto 2003)
$kk_{dswe} = 0.05$	(Ciliberto 2003)	$J_{awee} = 0.05$	(Ciliberto 2003)	$J_{iwee} = 0.0098$	this study
$kk_{swe} = 2$	(Ciliberto 2003)	$kkk_{swe} = 0.01$	(Ciliberto 2003)	$kkkk_{swe} = 0.2$	(Ciliberto 2003)
$k_{dHsl1Hsl7} = 4.93$	this study	$kk_{Hsl1Hsl7} = .001$	this study	$kkk_{Hsl1Hsl7} = 0.4$	this study
$kkd_{Hsl1Hsl7} = 0.001$	this study	$k_{dHog1mass} = 2.23$	this study	$Vi_{mih} = 0.3$	(Ciliberto 2003)
$Va_{mih} = 1$	(Ciliberto 2003)	$Ji_{mih} = 0.1$	(Ciliberto 2003)	$Ja_{mih} = 0.1$	(Ciliberto 2003)
$B_0 = 0.054$	(Chen 2004)	$kk_{mih1} = 5$	(Ciliberto 2003)	$kkk_{mih1} = 0.5$	(Ciliberto 2003)
$k_{hpkpc1} = 14$	this study	$k_{hkpc1} = 14$	this study	$kk_{ash1} = 3.2$	this study
$kkk_{ash1} = 0.2$	this study	$k_{Hog1Cln3} = 20$	this study	$n_{Hog1Cln3} = 2$	this study
$k_{dib5} = 0.06$	(Chen 2004)	$[Mih1]_T = 1$	(Ciliberto 2003)	$[Tem1]_T = 1$	(Ciliberto 2003)
$[Cdc15]_T = 1$	(Ghaemmaghami 2003)	$[Esp1]_T = 1$	(Ghaemmaghami 2003)	$kk_{ash} = 0.8$	this study
$J_{n3} = 6$					

$k_{mad2} = 8$ (for $[ORI] > 1$ and $[SPN] < 1$) or 0.01 (otherwise)
$k_{bub2} = 1$ (for $[ORI] > 1$ and $[SPN] < 1$) or 0.2 (otherwise)
$k_{lte1} = 1$ (for $[SPN] > 1$ and $[Clb2] < K_{ez}$) or 0.2 (otherwise)

All other parameters are dimensionless

References

1. A. Amon, M. Tyers, B. Futcher, K. Nasmyth, Mechanisms that help the yeast cell cycle clock tick: G2 cyclins transcriptionally activate G2 cyclins and repress G1 cyclins. Cell **74**(6), 993–1007 (1993)
2. K.C. Chen, A. Csikasz-Nagy, B. Gyorffy, J. Val, B. Novak, J.J. Tyson, Kinetic analysis of a molecular model of the budding yeast cell cycle. Mol. Biol. Cell **11**(1), 369–391 (2000)
3. K.C. Chen, L. Calzone, A. Csikasz-nagy, F.R. Cross, B. Novak, J.J. Tyson, Integrative analysis of cell cycle control in budding yeast. Mol. Biol. Cell, **15**, 3841–3862 (2004)
4. A. Ciliberto, B. Novak, J.J. Tyson, Mathematical model of the morphogenesis checkpoint in budding yeast. J. Cell Biol. **163**(6), 1243–1254 (2003)
5. F.R. Cross, V. Archambault, M. Miller, M. Klovstad, Testing a mathematical model of the yeast cell cycle. Mol. Biol. Cell, **13**(1), 52–70 (2002)
6. L. Dirick, T. Böhm, K. Nasmyth, Roles and regulation of Cln-Cdc28 kinases at the start of the cell cycle of Saccharomyces cerevisiae. EMBO J. **14**(19), 4803–4813, (1995)
7. C.B. Epstein, F.R. Cross, Genes that can bypass the CLN requirement for *Saccharomyces cerevisiae* cell cycle START. Mol. Cell. Biol. **14**(3), 2041–2047 (1994)
8. X. Escoté, M. Zapater, J. Clotet, F. Posas, Hog1 mediates cell-cycle arrest in G1 phase by the dual targeting of Sic1. Nat. Cell Biol. **6**(10), 997–1002 (2004)
9. S. Ghaemmaghami, W.-K. Huh, K. Bower, R.W. Howson, A. Belle, N. Dephoure, E.K. O'Shea, J.S. Weissman, Global analysis of protein expression in yeast. Nature **425**(6959), 737–741 (2003)
10. S.R. Salama, K.B. Hendricks, J. Thorner, G1 cyclin degradation: the PEST motif of yeast Cln2 is necessary, but not sufficient, for rapid protein turnover. Mol. Cell. Biol. **14**(12), 7953–7966 (1994)
11. E. Schwob, T. Böhm, M.D. Mendenhall, K. Nasmyth, The B-type cyclin kinase inhibitor p40SIC1 controls the G1 to S transition in *S. cerevisiae*. Cell **79**(20), 233–244 (1994)
12. T.T. Su, P.J. Follette, P.H. O'Farrell, Qualifying for the license to replicate. Cell **81**(6), 825–828 (1995)
13. W. Zachariae, K. Nasmyth, Whose end is destruction: Cell division and the anaphase-promoting complex. Genes Dev. **13**(16), 2039–2058 (1999)

Appendix B
Effect of Methods of Update on Existence of Fixed Points

In this Appendix we prove that the method of update does not alter the existence and number of fixed points. The Definitions and Lemma B are taken from the monograph by Robert [1].

The finite but large set of states of the Boolean network is denoted by $\mathbf{X}$, and F is a Boolean map from $\mathbf{X}$ to itself. We are interested in studying the dynamical behaviour of successively applying F to the elements of $\mathbf{X}$. Given an arbitrary initial state $\mathbf{x}^0 \in \mathbf{X}$, the sequence of states, i.e. the time evolution is given by

$$\mathbf{x}^{p+1} = F(\mathbf{x}^p) \quad (p = 0, 1, 2, \ldots). \tag{B.1}$$

Since $\mathbf{X}$ is a finite set, the sequence of state transitions can have two possible types of dynamical behaviour:

(i) Either the sequence converges to a stationary state $\mathbf{y}$ after a finite number of iterations; at this fixed point of F whereas $F(\mathbf{y}) = \mathbf{y}$.
(ii) Or the sequence repeats after a certain number of steps. The states of that recurrent sequence form a cycle $\{\mathbf{y}_1, \mathbf{y}_2, \ldots, \mathbf{y}_k\}$, which is defined by $\{\mathbf{y}_1 = F(\mathbf{y}_k), \mathbf{y}_2 = F(\mathbf{y}_1), \ldots, \mathbf{y}_k = F(\mathbf{y}_{k-1})\}$ [1].

Definition The iteration given by Eq. (B.1) is a *synchronous update* or *parallel mode of operation*. At each step all nodes are updated simultaneously.

Definition A *serial mode of operation* or *asynchronous update* is an iteration in which not all of the elements are updated at the same time. Starting from an arbitrary condition $\mathbf{x}^p$, one element is updated first, then the second, considering the effects of the changes to the first element, and so on. This leads to the system

$$x_1^{p+1} = f_1(x_1^p, x_2^p, \ldots, x_n^p)$$
$$x_2^{p+1} = f_2(x_1^{p+1}, x_2^p, \ldots, x_n^p)$$
$$\vdots \tag{B.2}$$

E. Radmaneshfar, *Mathematical Modelling of the Cell Cycle Stress Response*,
Springer Theses, DOI: 10.1007/978-3-319-00744-1,
© Springer International Publishing Switzerland 2014

$$x_n^{p+1} = f_n(x_1^{p+1}, x_2^{p+1}, \ldots, x_{n-1}^{p+1}, x_n^p),$$

where $x = (x_1 \ x_2 \ \ldots \ x_n)$ and $p \in \{0, 1, 2, \ldots\}$.

Definition The operator (Boolean function) F_i is defined by

$$F_i(x) = \begin{bmatrix} x_1 \\ \vdots \\ f_i(x_1 \ \cdots \ x_n) \\ \vdots \\ x_n \end{bmatrix}. \tag{B.3}$$

Lemma B.1 (Theorem 2 in [1]) *Every asynchronous update is a method of successive approximations of a Boolean function* $G : \mathbf{X} \to \mathbf{X}$, *called associated asynchronous operator for* $F : \mathbf{X} \to \mathbf{X}$.

$$G = F_n \circ \cdots \circ F_2 \circ F_1. \tag{B.4}$$

Proof. We will successively apply the asynchronous update to all $\mathbf{x} = (x_1 \ \cdots \ x_n) \in \mathbf{X}$:

$$g_1(x_1 \ \cdots \ x_n) = f_1(x_1 \ \cdots \ x_n)$$
$$g_2(x_1 \ \cdots \ x_n) = f_2(g_1(x) \ \cdots \ x_n)$$
$$\vdots$$
$$g_i(x_1 \ \cdots \ x_n) = f_i(g_1(x) \ \cdots \ g_{i-1}(x) \ x_i \ \cdots \ x_n)$$
$$\vdots$$
$$g_n(x_1 \ \cdots \ x_n) = f_n(g_1(x) \ \cdots \ g_{n-1}(x) \ x_n). \tag{B.5}$$

Hence, the final states, after successive application of the asynchronous update p times, can be obtained by directly applying the corresponding Boolean function G

$$\mathbf{x}^{p+1} = G(\mathbf{x}^p) \ (p = 0, 1, 2, \ldots). \tag{B.6}$$

$\square$

Corollary B.2 The sets of fixed points for the synchronous and asynchronous updates are identical.

Proof. This corollary is a direct consequence of lemma B. For every asynchronous map We can find a Boolean function G which is a sequential composition of F_i Eq. (B.4). Hence, if $\mathbf{y}$ is a fixed point of F, i.e. $F(\mathbf{y}) = \mathbf{y}$, this implies that $\mathbf{y}$ is also is

Appendix B: Effect of Methods of Update on Existence of Fixed Points   109

a fixed point of $G(\mathbf{y}) = F_n \circ \cdots \circ F_2 \circ F_1(\mathbf{y}) = \mathbf{y}$. Moreover, if $\mathbf{y}$ is a fixed point of G, then $G(\mathbf{y}) = \mathbf{y}$. According to Eq. (B.5), we have

$$g_1(y_1 \cdots y_n) = f_1(y_1 \cdots y_n) = y_1$$
$$g_2(y_1 \cdots y_n) = f_2(g_1(y) \cdots y_n) = y_2$$
$$\vdots$$
$$g_i(y_1 \cdots y_n) = f_i(g_1(y) \cdots g_{i-1}(y) \; y_i \cdots y_n) = y_i$$
$$\vdots$$
$$g_n(y_1 \cdots y_n) = f_n(g_1(y) \cdots g_{n-1}(y) \; y_n) = y_n.$$

$$(B.7)$$

Hence, $g_1(y_1, \ldots, y_n) = f_1(y_1 \cdots y_n) = y_1$. Therefore y_1 is a fixed point of F_1. With the same argument we can show that y_2 is the fixed point of F_2 and and argue iteratively for all F_n. As a result, $\mathbf{y} = (y_1 \cdots y_n)$ is a fixed point of F. Thus, F and G have the same set of fixed points. The synchronous or asynchronous application of the Boolean function does not change the existence and the set of fixed points.

Furthermore, the set of fixed points does not change if instead of updating the states purely synchronously or asynchronously we use a combination of both. The argument is analogous to the one above.

Reference

1. F. Robert, Discrete iterations: a metric study. Springer Series in Computational Mathematics (1986)

Printed by Publishers' Graphics LLC
LMO131023.15.14.32